Texts and Monographs in Computer Science

Editor

David Gries

Advisory Board
F.L. Bauer
S.D. Brookes
C.E. Leiserson
F.B. Schneider
M. Sipser

Texts and Monographs in Computer Science

(continued after index)

Larch: Languages and Tools for Formal Specification

John V. Guttag
James J. Horning

with S.J. Garland, K.D. Jones,
A. Modet, and J.M. Wing

With 76 Illustrations

Springer-Verlag
New York Berlin Heidelberg London Paris
Tokyo Hong Kong Barcelona Budapest

John V. Guttag
MIT Laboratory for Computer Science
545 Technology Square
Cambridge, MA 02139
USA

James J. Horning
Digital Equipment Corporation
 Systems Research Center
130 Lytton Avenue
Palo Alto, CA 94301-1044
USA

Series Editor:
David Gries
Department of Computer Science
Cornell University
Upson Hall
Ithaca, NY 14853-7501
USA

QA
76
. 6
G88
1993

Cover illustration: Variations on the Larch logo.

Library of Congress Cataloging-in-Publication Data
Guttag, J. V. (John V.)
    Larch : languages and tools for formal specification / John V.
 Guttag, James J. Horning.
      p.   cm.
    Includes bibliographical references.
    ISBN 0-387-94006-5 (U.S.)
    1. Electronic digital computers—Programming.   2. Computer
 software—Development.   3. Larch (Computer program language)
 I. Horning, James.   II. Title.
 QA76.6.H66   1993
 005.1′2—dc20   92-44571

Printed on acid-free paper.

Production managed by Bill Imbornoni; manufacturing supervised by Vincent Scelta.
Photocomposed from the authors' LAT$_{E}$X files.
Printed and bound by R.R. Donnelley & Sons, Inc., Harrisonburg, VA.
Printed in the United States of America.

9 8 7 6 5 4 3 2 1

ISBN 0-387-94006-5 Springer-Verlag New York Berlin Heidelberg
ISBN 3-540-94006-5 Springer-Verlag Berlin Heidelberg New York

Preface

Building software often seems harder than it ought to be. It takes longer than expected, the software's functionality and performance are not as wonderful as hoped, and the software is not particularly malleable or easy to maintain. It does not have to be that way.

This book is about programming, and the role that formal specifications can play in making programming easier and programs better. The intended audience is practicing programmers and students in undergraduate or basic graduate courses in software engineering or formal methods. To make the book accessible to such an audience, we have not presumed that the reader has formal training in mathematics or computer science. We have, however, presumed some programming experience.

The roles of formal specifications

Designing software is largely a matter of combining, inventing, and planning the implementation of abstractions. The goal of design is to describe a set of modules that interact with one another in simple, well-defined ways. If this is achieved, people will be able to work independently on different modules, and yet the modules will fit together to accomplish the larger purpose. In addition, during program maintenance it will be possible to modify a module without affecting many others.

Abstractions are intangible. But they must somehow be captured and communicated. That is what specifications are for. Specification gives us a way to say what an abstraction is, independent of any of its implementations.

The specifications in this book are written in formal specification languages. We use formal languages because we know of no other way to make specifications simultaneously as precise, clear, and concise. Anyone who has attempted to write documentation for a subroutine library, drafted contracts, or studied the tax code, knows how difficult it is to achieve even precision in a natural language—let alone clarity and brevity.

Mistakes from many sources will crop up in specifications, just as they do in programs. A great advantage of formal specification is that tools can be used to help detect and isolate many of these mistakes.

Some programmers are intimidated by the mere idea of formal specifications, which they fear may be too "mathematical" for them to understand and use. Such fears are groundless. Anyone who can learn to use a programming language can learn to use a formal specification language. After all, programs themselves are formal texts. Programmers cannot escape from formality and mathematical precision, even if they want to.

Overview of the book

Chapter 1 discusses the use of formal specifications in program development, providing a context for the technical material that follows. Chapter 2 contains a very short introduction to the notation of mathematical logic. The chapter is aimed at those with no background in logic, and provides all the logic background needed to understand the remainder of the book.

The rest of the book is an in-depth look at Larch, our approach to the formal specification of program components.

Chapter 3 gives an overview of the Larch two-tiered approach to specification. Each Larch specification has components written in two languages: one that is designed for a specific programming language (a Larch interface language) and another that is independent of any programming language (LSL, the Larch Shared Language). It also introduces LP, a tool used to reason about specifications. The descriptions are all brief; details are reserved for later chapters.

The remaining chapters are relatively independent, and can be read in any order. Chapter 4 is a tutorial on LSL. It is not a reference manual, but it does cover all features of the language. Chapter 5 is an introduction to LCL, a Larch interface language for Standard C. It describes the basic structure and semantics of the language, and it presents an extended example—along with hints about how to use LCL to support a style of C programming that emphasizes abstraction. Chapter 6 is an introduction to LM3, a Larch interface language for Modula-3. Chapter 7 discusses how LP can be used to analyze and help debug specifications written in LSL. It contains a short review of LP's major features, but is not comprehensive. Chapter 8 presents a brief summary of what we believe to be the essence of Larch.

The book concludes with several appendices. Appendix A contains a handbook of LSL specifications. Appendix B contains C implementations of the abstractions specified in Chapter 5. Appendix C deals with Larch's customization of lexical conventions. Appendix D contains a bibliography on Larch, and tells how to get more information about Larch, including how to get some of the Larch tools.

Some history

This book has been a long time in the growing. The seed was planted by Steve Zilles on October 3, 1973. During a programming language workshop organized by Barbara Liskov, he presented three simple equations relating operations on sets, and argued that anything that could reasonably be called a set would satisfy these axioms, and that anything that satisfied these axioms could reasonably be called a set. We developed this idea, and showed that all computable functions over an abstract type could be defined algebraically using equations of a simple form, and considered the question of when such a specification constituted an adequate definition [40].

As early as 1974, we realized that a purely algebraic approach to specification was unlikely to be practical. At that time, we proposed a combination of algebraic and operational specifications which we referred to as "dyadic specification" [39].

By 1980 we had evolved the essence of the two-tiered style of specification used in this book [43], although that term was not introduced until 1983 [86]. An early version of the Larch Shared Language was described in 1983 [44]. The first reasonably comprehensive description of Larch was published in 1985 [50]. Many readers complained that the contemporaneous *Larch in Five Easy Pieces* [51] should have been called *Larch in Five Pieces of Varying Difficulty*. They were not wrong.

By 1990 some software tools supporting Larch were available, and we began using them to check and reason about specifications. There is now a substantial and growing collection of support tools. We used them extensively in preparing this book. All of the formal proofs presented have been checked using LP. With the exception of parts of the LM3 specifications, all specifications have been subjected to mechanical checking. This process did not guarantee that the specifications accurately capture our intent; it did serve to help us find and eliminate several errors.

In the spring of 1990, we decided that it was time to make information on Larch more widely available. We originally thought of an anthology. The editors we contacted encouraged us to prepare a book, but urged us to provide a more coherent and integrated presentation of the material. We decided to take their advice. Had our families known how much of our time this would take, they would surely have tried to talk us out of it. In any event, we apologize to Andrea, David, Jane, Mark, Michael, and Olga for all the attention that "The Book" stole from them.

Acknowledgments

An important role in the development of Larch has been played by the organizations that provided the funding necessary to keep the project alive for so long. DARPA, NSF, the Digital Equipment Corporation, and Xerox were all valued patrons. A special debt of gratitude is owed to Bob Taylor, who as Director of the Computer Science Laboratory at the Xerox Palo Alto Research Center and then as Director of Digital's Systems Research Center has been a consistent supporter and friend. He encouraged people in his laboratories to work on Larch, he encouraged and funded efforts to transfer Larch to other parts of Digital, and he made possible the close collaboration between us by facilitating numerous visits by John Guttag, first to PARC and then to SRC, and by Jim Horning to MIT.

During the almost two decades we have been working on formal specification, we have accumulated a large number of intellectual debts. To list everyone who contributed an idea or an apt criticism would be impractical.

Over the years, Larch and related topics have been discussed at many meetings of IFIP Working Group 2.3. These discussions helped us to clarify our thinking in a number of areas.

Our early work on formal specification was influenced by a variety of people in the Department of Computer Science and the Computer Systems Research Group at the University of Toronto. The high degree of interaction between the theory and systems groups there provided a conducive atmosphere for this kind of work.

In the mid-seventies, John Guttag went to work at USC (and USC-ISI) and Jim Horning at Xerox PARC. Co-workers and visitors at both of these places played a significant role in the development of the Larch Shared Language and in helping us to understand the importance of support tools.

The most influential set of people have been our colleagues at Digital (particularly SRC) and at MIT's Laboratory for Computer Science (particularly members of the Systematic Program Development Group). They have encouraged our research and provided valuable technical feedback. Without their help, Larch would not exist. A few of these colleagues made particularly notable contributions. Jim Saxe's relentless criticism and creative suggestions contributed enormously to the development of both LSL and LP. Gary Feldman, Bill McKeeman, Yang Meng Tan, and Joe Wild contributed greatly to the design of LCL, as well as building and maintaining the LCL Checker. Greg Nelson provided the formal underpinnings on which the design of LM3 rests.

Our four co-authors played vital roles in the development of this book. They have worked with us on so many versions of the material in this book that we have not tried to record which words were whose. Steve Garland was a principal author of Chapters 2 and 7 and a vital contributor to the design of LSL, LCL, and LP. He also developed the majority of the software used to check and reason about the specifications appearing in this book. But for Steve, Larch would be a paper tiger. Kevin Jones was a principal designer of LM3 and provided much of the material in Chapter 6. Andrés Modet played a major role in the design and documentation of LSL. Jeannette Wing designed the first Larch interface language (Larch/CLU) and has been a vital contributor to almost all aspects of Larch ever since.

Many other people have helped in the preparation of this book. William Ang helped with the design of the artwork on the cover. Leslie Lamport provided a Larch style for LaTeX that made our life immeasurably easier. Manfred Broy, Daniel Jackson, Eric Muller, Sue Owicki, Fred Schneider, Mark Vandevoorde, and several anonymous reviewers provided extensive and helpful comments on various drafts. Cynthia Hibbard carefully edited the series of technical reports that led to this book. Judith Blount helped us to assemble and check the list of references. Jane Horning and Mary-Claire Van Leunen helped organize the index.

Finally, we wish to thank the Palo Alto Police Department for providing perspective. In August, a draft of this book was in a car that was stolen. Several days later the police recovered the car. When asked if any of the contents of the car had been recovered, they replied "Nothing of value." The thieves had removed everything from the car, except the manuscript.

J.V. Guttag and J.J. Horning
October 1992

Contents

Chapter 1

Specifications in Program Development

This book is about formal specification of programs and components of programs. We are interested in using specifications to help in the production and maintenance of high quality software.

We begin this chapter with a few remarks about programming and the role of abstraction. We then move on to discuss how specifications fit into the picture.

1.1 Programming with abstractions

Building a software system is almost entirely a design activity. Unfortunately, software is usually designed badly or barely designed at all. A symptom of negligence during design is the number of software projects that are seriously behind schedule, despite having had design phases that were "completed" right on schedule [10]. In practice, design is the phase of a software project that is declared "complete" when circumstances require it. Part of the problem is that there are few objective criteria for evaluating the quality and completeness of designs. Another part is the elapsed time between producing a design and getting feedback from the implementation process.

This book describes how formal specifications can be used effectively to structure and control the design process and to document the results.

The key to structuring and controlling the design process is, as Machiavelli said, "*Divide et impera.*" Regrettably, he was not clear about how to apply this stratagem to software development.

Two primary tools for dividing a problem are *decomposition* and *abstraction*. A good decomposition factors a problem into subproblems that:

- are all at the same level of detail,

- can be solved independently, and

- have solutions that can be combined to solve the original problem.

```
int sqrt(int x) {
  requires x ≥ 0;
  modifies nothing;
  ensures ∀ i: int
    ( abs(x - (result*result)) ≤ abs(x - (i*i)) );
}
```

FIGURE 1.1. A specification of an integer square Root procedure

The last criterion is the hardest to satisfy. This is where abstraction comes in. Abstraction involves ignoring details that are irrelevant for some purpose. It facilitates decomposition by making it possible to focus temporarily on simpler problems.

Consider, for example, the problem of designing a program to compile a source language, say Modula-3, to a target language, say Alpha machine code. Much of the compiler can be designed without paying attention to many of the details of either Modula-3 or the Alpha architecture. One might well begin by abstracting to the problem of compiling a source language with a deterministic context-free grammar to a reduced instruction (RISC) set target language. One might then choose to model the compiler's design on the design of other compilers that solve the same abstract problem, e.g., to decompose the problem into the separate problems of writing a scanner, a parser, a static semantic checker, and several code generation and optimization phases.

This paradigm of abstracting and then decomposing is typical of the program design process. Two important abstraction mechanisms are used: abstraction by parameterization and abstraction by specification.

Abstraction by parameterization allows a single program text to represent a potentially infinite set of computations or types. For example, the C code

```
      int twice(int x) {return x + x;}
```

denotes a procedure that can be used to double any integer.

Abstraction by specification allows a single text to represent a potentially infinite set of programs. For example, the specification in Figure 1.1 describes any procedure that, given an appropriate argument, computes an integer approximation to its square root. Notice that it specifies the required result, not any particular algorithm for achieving it. Notice also that it does not describe the result completely. For example, it does not constrain the result to be positive.

For the most part, software design is the process of inventing and combining abstractions and planning their implementation.

There are several reasons why it is better to think about combining abstractions than to think about combining their implementations:

- Abstractions are easier to understand than implementations, so combining abstractions is less work.

- Relying only on properties of the abstractions makes software easier to maintain, because it is clear what properties must be preserved when an implementation is changed.

- Because an abstraction can have several implementations with different performance properties, it can be used in various contexts with different performance requirements. Any implementation can be replaced by another during performance tuning without affecting correctness.

The key to good software design is inventing appropriate abstractions around which to structure the software. Bad programmers typically don't even try to invent abstractions. Mediocre programmers invent abstractions sufficient to solve the current problem. Great programmers invent elegant abstractions that get used again and again.

1.2   Finding abstractions

Structure is sometimes identified with hierarchy; hierarchical decomposition is sometimes preached as the only "structured" programming method. The problem with hierarchical decomposition is that, as the hierarchy gets deeper, it leads to highly specialized components that assume a great deal of context. This decreases the likelihood that components will be useful elsewhere—either in the current system or in software that is built later. A relatively flat structure usually encourages more reuse.

Important boundaries in the software should correspond to stable boundaries in the problem domain. Such correspondence makes it more likely that when customers ask for a small change in the observed behavior of the software, the change can be accomplished by a small change to the implementation. Stable boundaries in the problem domain frequently involve data types, rather than individual operations, because the kinds of objects that long-lived software manipulates tend to change more slowly than the operations performed on those objects. This leads to a style of programming in which data abstraction plays a prominent role.

A *data type* (data abstraction) is best thought of as a collection of related operations that manipulate a collection of related values [68]. For example, one should think of the type integer as providing operations, such as 0 and +, rather than as an array of 32 (or perhaps 64) bits, whose high-order bit is interpreted as its sign. Similarly, one should think of the type bond as a collection of operations such as get_coupon_rate and get_current_yield rather than as a record containing various fields.

An *abstract type* is a type that is presented to a client in terms of its specification, rather than its implementation. To implement an abstract type, one selects a *representation* (i.e., a storage structure and an interpretation that says how values of the type are represented) and implements the type's operations in terms of that representation. Clients of an abstract type invoke its operations, rather than directly accessing its representation. When the representation is changed, programs that use the type may have to be recompiled, but they needn't be rewritten.[1]

Even in languages, such as C, that provide no direct support for abstract types, there is a style of programming in which abstract types play a prominent role. Programmers rely on conventions to ensure that the implementation of an abstract type can be changed without affecting the correctness of software that uses the abstract type. The key restriction is that programs never directly access the representation of an abstract value. All access is through the operations (procedures and functions) provided in its interface.

1.3   The many roles of specification

Abstractions are intangible. But they must somehow be captured and communicated. Specification gives us a way to say what an abstraction is, independent of any of its implementations. Writing specifications can serve to clarify and deepen designers' understanding of whatever they are specifying, by focusing attention on possible inconsistencies, lacunae, and ambiguities.

Once written, specifications are helpful to implementors, testers, and

---

[1]For a more comprehensive discussion of the role of data abstraction in programming, see [63].

maintainers. Specifications provide "logical firewalls" by documenting mutual obligations. Implementors are to write software that meets its specification. Clients, i.e., writers of programs that use the software interface, are to rely only on properties of the software that are guaranteed by its specification.

During module testing and quality assurance, specifications provide information that can be used to generate test data, build stubs, and analyze information flow. During system integration, specifications reduce the number and severity of interface problems by reducing the number of implicit assumptions. Finally, specifications aid in maintenance by recording the properties that must be preserved and by delimiting the changes that might affect clients.

All of these virtues can be attributed to the information hiding provided by specifications. Specification makes it possible to completely hide the implementation of an abstraction from its clients, and to completely hide the uses made by clients from the implementor [70].

1.4   Styles of specification

A good specification should be tight enough to rule out implementations that are not acceptable. It should also be loose enough to allow the most desirable (i.e., efficient and elegant) implementations. A specification that fails to rule out undesired "solutions" is not sufficiently constraining; one that places unnecessary constraints on implementations is guilty of *implementation bias*.

A *definitional specification* explicitly lists properties that implementations must exhibit. The specification in Figure 1.1 is definitional. An *operational specification* gives one recipe that has the required properties, instead of describing them directly. Figure 1.2 contains an operational specification of a square root procedure. It looks suspiciously like a program—it defines a function by showing how to compute it. In fact, every program can be viewed as a specification. The converse is not true: many specifications are not programs. Programs have to be executable, but specifications don't. This freedom can often be exploited to make specifications simpler and clearer.

There are strong arguments in favor of both the operational and definitional styles of specification. The strength of operational specification lies in its similarity to programming. This reduces the time required for programmers to learn to use specifications. Some operational specifications

```
int sqrt(int x)
  requires x ≥ 0
  effects
    i = 0;
    while i*i < x
      i = i + 1 end
    if abs(i*i - x) > abs((i - 1) * (i - 1) - x)
      then return i - 1
      else return i
```

FIGURE 1.2. An Operational Specification of Integer Square Root

are directly executable. By executing specifications as "rapid prototypes," specifiers and their clients can get quick feedback about the software system being specified.

On the other hand, definitional specifications are not bound by the constraint of constructivity. They are often shorter and clearer than operational specifications. They are also easier to modularize, because properties can be stated separately and then combined. Because definitional specifications are so different from programs, they provide a distinct viewpoint on systems that is frequently helpful.

It is often difficult to determine from an operational specification which properties are necessary parts of the thing being specified and which are unimportant. The specification in Figure 1.2, for example, allows fewer implementations than the specification in Figure 1.1. An implementation is certainly not obliged to use the simple, but horribly inefficient, specification algorithm, but it must compute the same result, and therefore must not return a negative number. This constraint might be essential in some contexts and insignificant in others. Figure 1.2 does not say, and cannot easily be modified to say, whether the sign of the result matters. Figure 1.1, on the other hand, can easily be strengthened to specify the sign if that is important.

1.5   Formal specifications

The specifications in this book are written in formal specification languages. A formal specification language provides:

- a *syntactic domain*—the notation in which the specifications are written,

- a *semantic domain*—a universe of things that may be specified, and

- a *satisfaction relation* saying which things in the semantic domain satisfy (implement) which specifications in the syntactic domain.

We use formal languages because it seems to be the easiest way to write specifications that are simultaneously precise, clear, and concise. This is hardly surprising. It is no accident that such diverse activities as chemistry, chess, knitting, and music all have their own formal notations.

Mistakes from many sources will crop up in specifications, just as they do in programs. A great advantage of formal specification is that tools can be used to help detect and isolate many of these mistakes. Anyone who has used a strongly typed programming language knows that even something as simple as a syntax and type checker is invaluable. Comparable checking and diagnosis of formal specifications is easy and worthwhile, but we can do even better. Various kinds of formal specifications can be checked more thoroughly by tools that help explore the consequences of design decisions, detect logical inconsistencies, simulate execution, execute symbolically, prove the correctness of implementation steps (refinements), etc.

Are formal specifications too "mathematical" to be used by typical programmers? No. Anyone who can learn to read and write programs can learn to read and write formal specifications. After all, each programming language is a formal language.

Chapter 2

A Little Bit of Logic

This chapter contains all the logic one needs to know to understand Larch.

The mathematical formalism underlying the Larch family of languages is multisorted first-order logic with equality. We use a few notations and basic concepts from this logic quite freely in the rest of the book. If you are already familiar with logic, you should scan this chapter quickly to see which of the many "standard" logical notations we have adopted. If you have no acquaintance with logic, don't worry. This is a brief chapter, and the parts of logic that we present are really quite simple—almost as simple as basic arithmetic and much simpler than common programming languages. If you want a fuller treatment of logic, you should consult one of the many textbooks available, but there is no reason to do so before continuing in this book.

To help the your intuition, we point out programming analogs of some of the logical concepts. However, these analogies should not be pushed too far; logic is not a programming language. We use logic to describe properties that objects might or might not have (e.g., to describe what it means to be the shortest path between two points in a graph), whereas we use programming languages to describe how to produce certain objects (e.g., to describe how to find a shortest path).

2.1   Basic logical concepts

A logical language consists of a set of *sorts* and *operators* (function symbols). Sorts are much like programming language types. An operator (e.g., +) stands for a map from tuples of values to values; its *signature* (e.g., Int, Int→Int) is a tuple of sorts for its arguments (its *domain sorts*, e.g., Int, Int) and a sort for its result (its *range sort*, e.g., Int). A *relational operator* is a binary operator with range sort Bool (e.g., <:E, E→Bool). Operators are much like identifiers for value-returning procedures in programming languages.

An *application* consists of an operator and a tuple of terms, each of which has the same sort as the corresponding domain sort for the operator. The sort of an application is the same as that of the operator's range sort.

Applications are much like procedure calls in programming languages.

An important special case is an operator whose signature has no domain sorts. We will write such applications without parentheses (e.g., `empty` rather than `empty()`). We refer to both the operator and its application as a *constant*.

The application of an infix operator may be written with the operator between the two operands (e.g., `x+y` rather than `+(x, y)`). For operators that are associative, such as `+`, we also allow more than two operands (e.g., `x+y+z` is equivalent to `(x+y)+z` and to `+(+(x, y), z)`).

A *variable* is an identifier standing for an arbitrary value of some sort. Logical variables are different from programming language variables because the value of a logical variable does not change over time.

A *term* is a variable, an application, or a parenthesized term.

An *equation* is a term of sort `Bool`, written as a pair of terms of the same sort, joined by the the *equality operator*, =.

A *predicate* (also called a *formula*) is a term of sort `Bool`. In order to determine whether a given predicate is true or false, we must know how to interpret the sorts and operators in the logical language. For example, `sqrt(5) = 2` is false if `sqrt` is interpreted as the square-root function over the real numbers and the constant operators 5 and 2 are interpreted as the real numbers five and two. Alternatively, the predicate is true if `sqrt` is interpreted as the greatest-integer-less-than-or-equal-to-the-square-root function. So it only makes sense to talk about whether a predicate is true or false if we are given a *structure* (interpretation) that assigns

- a nonempty set of values to each sort, and

- a total function (that maps tuples of values of its domain sorts to values of its range sort) to each operator.

Most logics come with a set of operators whose meanings are fixed *a priori*, for example, the equality operator for each sort. Others are the *propositional connectives* ⇔ (if and only if), ¬ (not), ∧ (and), ∨ (or), and ⇒ (implies).

First-order logic provides several ways to form predicates. We describe these, as well as what it means for each kind of predicate to be true in a given structure under a given assignment of values to its variables.

- As mentioned above, an *equation* is a predicate consisting of a pair of terms of the same sort, joined by the equality operator, =. It is true if its two operands have the same value in the given structure under the given assignment of values to variables. The predicate x = y may be read as "x equals y." The propositional connective ⇔ has the same meaning as the equality operator for the sort Bool. The predicate P ⇔ Q may be read as "P if and only if Q."

- A *negation* is a predicate preceded by the negation operator, ¬. It is true if the operand of ¬ is false. The predicate ¬P may be read as "not P."

- A *conjunction* is a pair of predicates joined by the conjunction connective, ∧. A conjunction is true if both its operands are true. The predicate P ∧ Q may be read as "both P and Q."

- A *disjunction* is a pair of predicates joined by the disjunction connective, ∨. A disjunction is true if at least one of its operands is true. The predicate P ∨ Q may be read as "either P or Q or both."

- An *implication* is a pair of predicates joined by the implication connective, ⇒. An implication is true if its left operand is false or its right operand is true. Therefore, P ⇒ Q has the same meaning as ¬P ∨ Q. The predicate P ⇒ Q may be read as "P implies Q" or "if P then Q."

- A *binding* is a predicate preceded by a variable and its sort. All occurrences of the variable in the predicate are said to be *bound* (and to have that sort). The binding is said to have *captured* the variable it binds. A variable is *free* in a predicate if there are any instances of it anywhere in the predicate that are not bound.

- A *quantified predicate* is a binding preceded by either the *existential quantifier*, ∃, or the *universal quantifier*, ∀. Bindings are only allowed immediately following quantifiers. The binding ∀x : S may be read as "for all x of sort S."

  - A *witness* for a bound variable is a value that makes the predicate in its binding true, in a structure under a given assignment, when the assignment is modified to assign the witness to the bound variable.

  - An existentially quantified predicate is true if there is at least one witness for its bound variable. The predicate ∃x : S (P) may be read as "there exists an x of sort S such that P."

– A universally quantified predicate is true if the predicate in its binding is true for all values of its bound variable. The predicate $\forall x : S\ (P)$ may be read as "for all x of sort S, P."

If a predicate is true in all structures under all assignments to its free variables, it is said to be *valid* or a *tautology*. If there exists a structure and an assignment to its free variables under which it is true, it is said to be *satisfiable*.

A *sentence* is a predicate with no free variables. By convention, we consider a free-standing predicate with free variables as standing for the sentence obtained by universally quantifying its free variables at the outermost level. Since the truth of a predicate in a structure depends only on the values assigned to its free variables, and since a sentence contains no free variables, we talk about a sentence being true in a structure, rather than in a structure under an assignment.

When a sentence is true in a structure, we say that the structure is a *model* of that sentence. Similarly, when each member of a set of sentences is true in a structure, we say that the structure is a model of that set. Consider, for example, a language with a single non-Bool sort, E, with one operator, the binary relation <, and with three variables x, y, and z of sort E. Any structure that is a model of the two sentences

$$\forall\ x : E\ \neg (x < x)$$

$$\forall\ x : E\ \forall\ y : E\ \forall\ z : E\ ((\ x < y \wedge y < z) \Rightarrow x < z)$$

is commonly known as a *strict partial order*, and we call these sentences *axioms* for strict partial orders.

A sentence S is a *logical consequence* of a set T of sentences if every model of T is also a model of S. For example, the sentence

$$\forall\ x : E\ \forall\ y : E\ \neg (x < y \wedge y < x)$$

is a consequence of the axioms for strict partial orders, because it is true in all strict partial orders.

A set of sentences is *closed* under logical consequence if it contains all its logical consequences. A *theory* is a set of sentences closed under logical consequence. For example, the theory of strict partial orders is the set of all consequences of the axioms for strict partial orders; equivalently, it is the set of sentences true in all strict partial orders.

A theory is *complete* if for every sentence S, either S or ¬S is in the theory. Most of the time, we find ourselves dealing with incomplete theories. For example, there is no computable set of sentences whose logical consequences are exactly the sentences true about the natural numbers under the usual operations of addition and multiplication.

A set of sentences is *consistent* if it has a model. It is easy to show that a sentence S is a consequence of a set T of sentences if and only if T $\cup$ {¬S} is inconsistent. Likewise, a theory is consistent if and only if it does not contain a *contradiction*, that is, the sentence `true = false`.

2.2   Proof and consequences

In the preceding section, we provided a semantic description of what it means for a sentence S to be a logical consequence of a set of sentences T, namely that every model of T also be a model of S. Unfortunately, this definition does not provide a practical means for determining when S is a logical consequence of T. For example, T may have infinitely many models, some of its models may have infinitely many elements, etc.

Fortunately, there is a syntactic characterization of what it means for S to be a logical consequence of T. A formal *deduction system* consists of a set of sentences (called *logical axioms*) together with a set of functions (called *deduction rules*) that map finite sets of sentences (the *premises* of a deduction) to a single sentence (its *conclusion*). For example, the deduction rule

```
    P, P ⇒ Q
    -----------
         Q
```

states that Q can be deduced from the premises P and P $\Rightarrow$ Q.

A *proof* based on a set T of sentences is a finite sequence of sentences each of which is either a logical axiom, a member of T, or the conclusion of a deduction rule applied to a set of sentences occurring earlier in the proof. A sentence S is a *theorem* of T if it occurs in some proof based on T.

There are three properties that a good formal system of deduction should possess:

- It should not allow any spurious proofs. A system is *sound* if, for any T, every theorem of T is really a logical consequence of T.

- It should provide enough proofs. A system is *complete* if, for any T, every logical consequence of T is also a theorem of T.

- It should be possible to recognize what is a proof and what is not. A system is *effective* if, for any computable set T of sentences, the set of proofs based on T is also computable.

There are several sound, complete, and effective formal systems of deduction for first-order logic. For most of this book, the mere existence of good formal systems of deduction is all that counts. The choice of a particular system, or the details of that system (which we refer to as "the usual rules of first-order logic"), do not really matter. What matters is that the system is sound (because we do not want to prove anything that isn't true) and effective (because we want to know when we have a proof). Completeness of a deductive system matters less, since we often find ourselves dealing with incomplete theories. Of course, the system of deduction used in LP, Chapter 7, is sound and effective.

This concludes our whirlwind introduction to the vocabulary and notation of mathematical logic used in the remainder of this book. We rely primarily on the predicate-forming operators described on pages 9–11.

Chapter 3

An Introduction to Larch

We begin this chapter by describing the Larch approach to specification and illustrating it with a few small examples. Our intent is to give you a taste of Larch. Details are reserved for later chapters. We then discuss LP, the Larch proof assistant, a tool that supports all the Larch languages. Again, we give only a taste. Finally, we discuss the lexical and typographical conventions used for preparing and presenting the Larch specifications in this book.

3.1   Two-tiered specifications

The Larch family of languages supports a *two-tiered*, definitional style of specification. Each specification has components written in two languages: one language that is designed for a specific programming language and another language that is independent of any programming language. The former kind are *Larch interface languages*, and the latter is the *Larch Shared Language* (LSL).

Interface languages are used to specify the interfaces between program components. Each specification provides the information needed to use an interface. A critical part of each interface is how components communicate across the interface. Communication mechanisms differ from programming language to programming language. For example, some languages have mechanisms for signalling exceptional conditions, others do not. More subtle differences arise from the various parameter passing and storage allocation mechanisms used by different languages.

It is easier to be precise about communication when the interface specification language reflects the programming language. Specifications written in such interface languages are generally shorter than those written in a "universal" interface language. They are also clearer to programmers who use components and to programmers who implement them.

Each interface language deals with what can be observed by client programs written in a particular programming language. It provides a way to write assertions about program states, and it incorporates programming-language-specific notations for features such as side effects, exception

```
uses TaskQueue;
mutable type queue;
immutable type task;

task *getTask(queue q) {
  modifies q;
  ensures
    if isEmpty(q^)
      then result = NIL ∧ unchanged(q)
      else (*result)' = first(q^) ∧ q' = tail(q^);
}
```

FIGURE 3.1. An LCL interface specification

handling, iterators, and concurrency. Its simplicity or complexity depends largely on the simplicity or complexity of its programming language.

Larch interface languages have been designed for a variety of programming languages. The two that are discussed in this book are for C and for Modula-3. Other interface languages have been designed for Ada [15, 37], CLU [86], C++ [60, 90, 92], ML [93], and Smalltalk [17]. There are also "generic" Larch interface languages that can be specialized for particular programming languages or used to specify interfaces between programs in different languages [16, 53, 61, 88].

Larch interface languages encourage a style of programming that emphasizes the use of abstractions, and each provides a mechanism for specifying abstract types. If its programming language provides direct support for abstract types (as Modula-3 does), the interface language facility is modeled on that of the programming language; if its programming language does not (as C does not), the facility is designed to be compatible with other aspects of the programming language.

Figure 3.1 contains a sample interface specification for a small fragment of a scheduler for an operating system. The specification is written in LCL (a Larch interface language for C, described in Chapter 5). This fragment introduces two abstract types and a procedure for selecting a task from a task queue. Briefly, * means pointer to (as in C), result refers to the value returned by the procedure, the symbol ^ is used to refer to the value in a location when the procedure is called, and the symbol ' to refer to its value when the procedure returns.

The specification of getTask is not self-contained. For example, looking only at this specification there is no way to know which task

```
TaskQueue: trait
  includes Nat
  task tuple of id: Nat, important: Bool
  introduces
    new: → queue
    __ ⊣ __: task, queue → queue
    isEmpty, hasImportant: queue → Bool
    first: queue → task
    tail: queue → queue
  asserts
    queue generated by new, ⊣
  ∀ t: task, q: queue
    isEmpty(new);
    ¬isEmpty(t ⊣ q);
    ¬hasImportant(new);
    hasImportant(t ⊣ q) ==
      t.important ∨ hasImportant(q);
    first(t ⊣ q) ==
      if t.important ∨ ¬hasImportant(q)
        then t else first(q);
    tail(t ⊣ q) ==
      if first(t ⊣ q) = t then q else t ⊣ tail(q)
```

FIGURE 3.2. LSL specification used by getTask

getTask selects. Is it the one that has been in q the longest? Is it is the one in q with the highest priority?

Interface specifications rely on definitions from *auxiliary specifications*, .written in LSL, to provide semantics for the primitive terms they use. Specifiers are not limited to a fixed set of notations, but can use LSL to define specialized vocabularies suitable for particular interface specifications or classes of specifications.

Figure 3.2 contains a portion of an LSL specification that specifies the operators used in the interface specification of getTask. Based on the information in this LSL specification, one can deduce that the task pointed to by the result of getTask is the most recently inserted important task, if such a task exists. Otherwise it is the most recently inserted task.

Many informal specifications have a structure similar to this. They implicitly rely on auxiliary specifications by describing an interface in terms of concepts with which readers are assumed to be familiar, such as sets, lists, coordinates, and windows. But they don't define these auxiliary concepts. Readers can misunderstand such specifications, unless their intuitive understanding exactly matches the specifier's. And there is no way to be sure that such intuitions do match. LSL specifications provide unambiguous mathematical definitions of the terms that appear in interface specifications.

Larch encourages a separation of concerns, with basic constructs in the LSL tier and programming details in the interface tier. We suggest that specifiers keep most of the complexity of specifications in the LSL tier for several reasons:

- LSL specifications are likely to be more reusable than interface specifications.

- LSL has a simpler underlying semantics than most programming languages (and hence than most interface languages), so specifiers are less likely to make mistakes, and any mistakes they do make are more easily found.

- It is easier to make and to check assertions about semantic properties of LSL specifications than about semantic properties of interface specifications.

Many programming errors are easily detected by running the program, that is, by testing it. While some Larch specifications can be executed, most of them cannot. The Larch style of specification emphasizes brevity and clarity rather than executability. To make it possible to validate specifications before implementing or executing them, Larch permits specifiers to make assertions about specifications that are intended to be redundant. These assertions can be checked mechanically. Several tools that assist specifiers in checking these assertions as they debug specifications are already in use, and others are under development.[1]

3.2   LSL, the Larch Shared Language

LSL specifications define two kinds of symbols, *operators* and *sorts*. The concepts of operator and sort are the same as those used in Chapter 2. They are similar to the programming language concepts of procedure and type, but it is important not to confuse these two sets of concepts. When discussing LSL specifications, we will consistently use the words "operator" and "sort." When talking about programming language constructs, we will use the words "procedure" (or "function," "routine," or "method," as appropriate) and "type." As discussed in Chapter 2, operators stand for total functions from tuples of values to values. Sorts stand for disjoint non-empty sets of values, and are used to indicate the domains and ranges of operators. In each interface language, "procedure" and "type" must mean what they mean in that programming language.

The *trait* is the basic unit of specification in LSL. A trait introduces some operators and specifies some of their properties. Sometimes the trait defines an abstract type. However, it is frequently useful to define a set of properties that does not fully characterize a type.

Figure 3.3 shows a trait that specifies a class of tables that store values in indexed places. It is similar to specifications in many "algebraic" specification languages.

The specification begins by *including* another trait, Integer. This specification, which can be found in the LSL handbook in Appendix A, page 163, supplies information about the operators +, 0, and 1, which are used in defining the operators introduced in Table.

The *introduces clause* declares a set of operators, each with its *signature* (the sorts of its domain and range). Signatures are used to sort-check terms in much the same way as procedure calls are type-checked in programming languages.

The *body* of the specification contains, following the reserved word

---

[1] See Appendix D for a list.

```
Table: trait
    includes Integer
    introduces
        new: → Tab
        add: Tab, Ind, Val → Tab
        __ ∈ __: Ind, Tab → Bool
        lookup: Tab, Ind → Val
        size: Tab → Int
    asserts ∀ i, i1: Ind, v: Val, t: Tab
        ¬ (i ∈ new);
        i ∈ add(t, i1, v) == i = i1 ∨ i ∈ t;
        lookup(add(t, i, v), i1) ==
            if i = i1 then v else lookup(t, i1);
        size(new) == 0;
        size(add(t, i, v)) ==
            if i ∈ t then size(t) else size(t) + 1
```

FIGURE 3.3. Table.lsl

`asserts`, equations between terms containing operators and variables.[2] The third equation resembles a recursive function definition, since the operator `lookup` appears on both the left and right sides. However, it merely states a relation that must hold among `lookup`, `add`, and the built-in operator `if__then__else__`; it does not fully define `lookup`. For example, it doesn't say anything about the value of the term `lookup(new, i)`.

The *theory* of a trait is the set of all logical consequences of its assertions. It is an infinite set of formulas in multisorted first-order logic with equality. It contains everything that logically follows from its assertions, but nothing else. The theory associated with `Table` contains equalities and disequalities that can be proved by substitution of equals for equals. LSL also provides two constructs for non-equational assertions that can be used to generate stronger (larger) theories. These important constructs are discussed in Chapter 4.

It is instructive to note some of the things that `Table` does *not* specify:

1. It does not say how tables are to be represented.

2. It does not give algorithms to manipulate tables.

3. It does not say what procedures are to be implemented to operate on tables.

4. It does not say what happens if one looks up an `Ind` that is not in a `Tab`.

The first two decisions are in the province of the implementation. The third and fourth are recorded in interface specifications.

3.3   Interface specifications

An interface specification defines an interface between program components, and is written in a programming-language-specific Larch interface language. Each specification must provide the information needed to use an interface and to write programs that implement it. At the core of each Larch interface language is a model of the state manipulated by the associated programming language.

---

[2]The equation connective in LSL, ==, has the same semantics as the equality symbol, =. It is used only to introduce another level of precedence into the language.

PROGRAM STATES

States are mappings from *locs* (abstract storage locations, also known as objects) to *values*. Each variable identifier has a type and is associated with a loc of that type. The major kinds of values that can be stored in locs are:

- *basic values*. These are mathematical constants, like the integer 3 and the letter A. Such values are independent of the state of any computation.

- *exposed types*. These are data structures that are fully described by the type constructors of the programming language (e.g., C's int * or Modula-3's ARRAY [1..10] OF INTEGER). The representation is visible to, and may be relied on by, clients.

- *abstract types*. As mentioned in Chapter 1, data types are best thought of as collections of related operations on collections of related values. Abstract types are used to hide representation information from clients.

Each interface language provides operators (e.g., ^ and ') that can be applied to locs to extract their values in the relevant states (usually the pre-state and the post-state of a procedure).

Each loc's type defines the kind of values it can map to in any state. Just as each loc has a unique type, each LSL term has a unique sort. To connect the two tiers in a Larch specification, there is a mapping from interface language types (including abstract types) to LSL sorts. Each type of basic value, exposed type, and abstract type is *based on* an LSL sort. Interface specifications are written using types and values. Properties of these values are defined in LSL, using operators on the corresponding sorts.

For each interface language, a standard LSL trait defines operators that can be applied to values of the sorts that the programming language's basic types and other exposed types are based on. Users familiar with the programming language will already have an intuitive understanding of these operators. Abstract types are typically based on sorts defined in traits supplied by specifiers.

PROCEDURE SPECIFICATIONS

The specification of each procedure in an interface can be studied, understood, and used without reference to the specifications of other procedures. A specification consists of a procedure header (declaring the types of its arguments and results) followed by a body of the form:

```
requires reqP
modifies modList
ensures  ensP
```

A specification places constraints on both clients and implementations of the procedure. The *requires clause* is used to state restrictions on the state, including the values of any parameters, at the time of any call. The *modifies* and *ensures clauses* place constraints on the procedure's behavior when it is called properly. They relate two states, the state when the procedure is called, the *pre-state*, and the state when it terminates, the *post-state*.

A requires clause refers only to values in the pre-state. An ensures clause may also refer to values in the post-state.

A modifies clause says what locs a procedure is allowed to change (its *target list*). It says that the procedure must not change the value of any locs visible to the client except for those in the target list. Any other loc must have the same value in the pre and post-states. If there is no modifies clause, then nothing may be changed.

For each call, it is the responsibility of the client to make the requires clause true in the pre-state. Having done that, the client may assume that:

- the procedure will terminate,

- changes will be limited to the locs in the target list, and

- the postcondition will be true on termination.

The client need not be concerned with how this happens.

The implementor of a procedure is entitled to assume that the precondition holds on entry, and is only responsible for the procedure's behavior if it is. A procedure's behavior is totally unconstrained if its precondition isn't satisfied, so it is good style to keep the requires clause weak. An omitted requires clause is equivalent to `requires true` (the weakest possible requirement).

TWO INTERFACE LANGUAGE EXAMPLES

Figure 3.4 contains a fragment of a specification written in LCL (a Larch interface language for Standard C). Figure 3.5 contains a fragment of a similar specification written in LM3 (a Larch interface language for Modula-3). They use the same `Table` trait of Figure 3.3. We present

```
mutable type table;
uses Table(table for Tab, char for Ind,
          char for Val, int for Int);
constant int maxTabSize;

table table_create(void) {
  ensures result' = new ∧ fresh(result);
  }
bool table_add(table t, char i, char c) {
  modifies t;
  ensures result = (size(t^) < maxTabSize ∨ i ∈ t^)
    ∧ (if result then t' = add(t^, i, c)
               else t' = t^);
  }
char table_read(table t, char i) {
  requires i ∈ t^;
  ensures result = lookup(t^, i);
  }
```

FIGURE 3.4. A Sample LCL Interface Specification

```
INTERFACE Table;
< * TRAITS Table(CHAR FOR Ind, CHAR FOR Val,
                INTEGER FOR Int) * >
  TYPE T <: OBJECT
    METHODS
      Add(i: CHAR; c: CHAR) RAISES {Full};
      Read(i: CHAR): CHAR;
    END;
  PROCEDURE Create( ): T;
  CONST MaxTabSize: INTEGER = 100;
  EXCEPTION Full;
< *
  FIELDS OF T
    val :   Tab;
  METHOD T.Add(i, c)
    MODIFIES SELF.val
    ENSURES SELF.val' = add(SELF.val, i, c)
      EXCEPT size(SELF.val) ≥ MaxTabSize
          ∧ ¬(i ∈ SELF.val)
        => RAISEVAL = Full ∧ UNCHANGED(ALL)
  METHOD T.Read(i)
    REQUIRES i ∈ SELF.val
    ENSURES RESULT = lookup(SELF.val, i)
  PROCEDURE Create
    ENSURES RESULT.val = new ∧ FRESH(RESULT)
* >
END Table.
```

FIGURE 3.5. A Sample LM3 Interface Specification

```
void choose(int x, int y) int z; {
  modifies z;
  ensures z' = x ∨ z' = y;
}
```

FIGURE 3.6. A specification of choose

these examples here simply to convey an impression of how programming language dependencies influence Larch interface languages. At this point, you should not be concerned with their exact meaning; the notations used are described in detail in Chapters 5 and 6.

3.4   Relating implementations to specifications

In this book we emphasize using specifications as a communication medium. Programmers are encouraged to become clients of well-specified abstractions that have been implemented by others. This book does not discuss the process of implementing specifications; there is already a copious literature on the subject.

One of the advantages of Larch's two-tiered approach to specification is that the relationship of implementations to specifications is relatively straightforward. Consider, for example, the LCL specification in Figure 3.6 and the C implementation in Figure 3.7.

The specification defines a relation between the program state when choose is called and the state when it returns. This relation contains all pairs of states $<pre, post>$ in which

- the states differ only in the value of the global variable z, and

- in *post* the value of z is that of one of the two arguments passed to choose.

The implementation also defines a relation on program states. This relation contains all pairs of states $<pre, post>$ in which

- the states differ only in the value of the variable z, and

- in *post* the value of z is the maximum of the two arguments passed to choose.

```
void choose(int x, int y) {
  if (x > y) z = x;
  else z = y;
}
```

FIGURE 3.7. An implementation of choose

We say that the implementation of choose in Figure 3.7 *satisfies* the specification in Figure 3.6—or is a *correct implementation* of Figure 3.6[3]— because the relation defined by the implementation is a subset of the relation defined by the specification. Every possible behavior that can be observed by a client of the implementation is permitted by the specification.

The definition of satisfaction we have just given is not directly useful. In practice, formal arguments about programs are not usually made by building and comparing relations. Instead, such proofs are usually done by pushing predicates through the program text, in ways that can be justified by appeal to the definition of satisfaction. A description of how to do this appears in the books [21, 36].

The notion of satisfaction is a bit more complicated for implementations of abstract types, because the implementor of an abstract type is working simultaneously at two levels of abstraction. To implement an abstract type, one chooses data structures to represent values of the type, then writes the procedures of the type in terms of that representation. However, since the specifications of those procedures are in terms of abstract values, one must be able to relate the representation data structures to the abstract values that they represent. This relation is an essential (but too often implicit) part of the implementation.

Figure 3.8 shows an implementation of the LCL specification in Figure 3.4. A value of the abstract type table is represented by a pointer to a struct containing two arrays and an integer. You need not look at the details of the code to understand the basic idea behind this implementation. Instead, you should consider the abstraction function and representation invariant.

The *abstraction function* is the bridge between the data structure used

---

[3]"Correct" is a dangerous word. It is not meaningful to say that an implementation is "correct" or "incorrect" without saying what specification it is claimed to satisfy. The technical sense of "correct" that is used in the formal methods community does not imply "good," or "useful," or even "not wrong," but merely "consistent with its specification."

```
#include "bool.h"
#define maxTabSize (10)

typedef struct {char ind[maxTabSize];
                char val[maxTabSize];
                int next;} tableRep;
typedef tableRep * table;

table table_create(void) {
  table t;
  t = (table) malloc(sizeof(tableRep));
  if (t == 0) {
    printf("Malloc returned null in table_create\n");
    exit(1);
  }
  t->next = 0;
  return t;
}
bool table_add(table t, char i, char c) {
  int j;
  for (j = 0; j < t->next; j++)
    if (t->ind[j] == i) {
      t->val[j] = c;
      return TRUE;
    }
  if (t->next == maxTabSize) return FALSE;
  t->val[t->next++] = c;
  return TRUE;
}
char table_read(table t, char i) {
  int j;
  for (j = 0; TRUE; j++)
    if (t->ind[j] == i) return t->val[j];
}
```

FIGURE 3.8. Implementing an abstract type

in the implementation of an abstract type and the abstract values being implemented. It maps each value of the representation type to a value of the abstract type. Here, we represent a `table` by a pointer, call it `t`, to a struct. If the triple `<ind, val, next>` contains the values of the fields of that struct in some state s, then we can define the abstract value represented by `t` in state s as `toTab(<ind, val, next>)`, where

```
toTab(<ind, val, next>) ==
  if next = 0 then empty
  else insert(toTab(<next - 1, ind, val>),
              ind[next], val[next])
```

Abstraction functions are often many-to-one. Here, for example, if `t->next = 0`, `t` represents the empty `table`, no matter what the contents of `t->ind` and `t->val`.

The typedefs in Figure 3.8 define a data structure sufficient to represent any value of type `table`. However, it is not the case that any value of that data structure represents a value of type `table`. In defining the abstraction function, we relied upon some implicit assumptions about which data structures were valid representations. For example, `toTab` is not defined when `t->next` is negative. A *representation invariant* is used to make such assumptions explicit. For this implementation, the representation invariant is

- The value of `next` lies between 0 and `maxTabSize`:

$$0 \leq \texttt{t->next} \ \wedge \ \texttt{t->next} \leq \texttt{maxTabSize}$$

- and no index may appear more than once in the fragment of `ind` that lies between 0 and `next`:

```
∀ i,j:int
  (0 ≤ i ∧ i < j ∧ j < t->next)
  ⇒ (t->ind)[i] ≠ (t->ind)[j]
```

To show that that this representation invariant holds, we use a proof technique called *data type induction*. Since `table` is an abstract type, we know that clients cannot directly access the data structure used to represent a `table`. Therefore, all values of type `table` that occur during program execution will have been generated by the functions specified in the interface. So to show that the invariant holds it suffices to show, reasoning from the code implementing the functions on `tables`, that

- the value returned by `table_create` satisfies the invariant (this is the basis step of the induction),

- whenever `table_add` is called, if the invariant holds for `t^` then the invariant will also hold for `t'`, and

- whenever `table_read` is called, if the invariant holds for `t^` then the invariant will also hold for `t'`.

A slightly different data type induction principle can be used to reason about clients of abstract types. To prove that a property holds for all instances of the type, i.e., that it is an *abstract invariant*, one inducts over all possible sequences of calls to the procedures that create or modify locs of the type. However, one reasons using the specifications of the procedures rather than their implementations. For example, to show that the `size(t)` is never greater than `maxTabSize` one shows that

- the specification of `table_create` implies that the size of the `table` returned is not greater than `maxTabSize`, and

- the specification of `table_add` combined with the hypothesis `t^` $\leq$ `maxTabSize` implies that `t'` $\leq$ `maxTabSize`.

Given the abstraction function, it is relatively easy to define what it means for the procedure implementations in Figure 3.8 to satisfy the specifications in Figure 3.4. For example, we say that the implementation of `table_read` satisfies its specification because the image under the abstraction function of the relation between pre and post-states defined by the implementation (i.e., what one gets by applying the abstraction function to all values of type `table` in the relation defined by the implementation) is a subset of the relation defined by the specification. Notice, by the way, that any argument that the implementation of `table_read` satisfies its specification will rely on both the `requires` clause of the specification and on the representation invariant.

3.5   LP, the Larch proof assistant

The discussions of LSL, LCL, and LM3 have alluded to tools supporting those languages. LP is a tool that is used to support all three. Chapter 7, which is about reasoning about LSL specifications, contains a brief description of LP. Here we give merely a glimpse of its use.

LP is a proof assistant for a subset of multisorted first-order logic with equality, the logic—described in Chapter 2—on which the Larch languages are based. It is designed to work efficiently on large problems and to be used by specifiers with relatively little experience with theorem proving. Its design and development have been motivated primarily by our work on LSL, but it also has other uses, for example, reasoning about circuit designs [75, 79], algorithms involving concurrency [25], data types [92], and algebraic systems [65].

LP is intended primarily as an interactive proof assistant or proof debugger, rather than as a fully automatic theorem prover. Its design is based on the assumption that initial attempts to state and prove conjectures usually fail. So LP is designed to carry out routine (but possibly lengthy) proof steps automatically and to provide useful information about why proofs fail. To keep users from being surprised and confused by its behavior, LP does not employ complicated heuristics for finding proofs automatically. It makes it easy for users to employ standard techniques such as proof by cases, by induction, or by contradiction, but the choice among such strategies is left to the user.

THE LIFE CYCLE OF PROOFS

Proving is similar to programming: proofs are designed, coded, debugged, and (sometimes) documented.

Before designing a proof it is necessary to formalize the things being reasoned about and the conjecture to be proved. The design of the proof proper starts with an outline of its structure, including key lemmas and methods of proof. The proof itself must be given in sufficient detail to be convincing. What it means to be convincing depends on who (or what) is to be convinced. Experience shows that humans are frequently convinced by unsound proofs, so we look for a mechanical "skeptic" that is just hard enough (but not too hard) to convince.

Once part of a proof has been coded, LP can be used to debug it. Proofs of interesting conjectures hardly ever succeed the first time. Sometimes the conjecture is wrong. Sometimes the formalization is incorrect or incomplete. Sometimes the proof strategy is flawed or not detailed enough. LP provides a variety of facilities that can be used to understand the problem when an attempted proof fails.

While debugging proofs, users frequently reformulate axioms and conjectures. After any change in the axiomatization, it is necessary to recheck not only the conjecture whose proof attempt uncovered the

```
Nat: trait
  includes AC(+, Nat)
  introduces
    0: → Nat
    s: Nat → Nat
    __ < __: Nat, Nat → Bool
  asserts
    Nat generated by 0, s
    ∀ i, j, k: Nat
      i + 0 == i;
      i + s(j) == s(i + j);
      ¬(i < 0);
      0 < s(i);
      s(i) < s(j) == i < j
    implies ∀ i, j, k: Nat
      i < j ⇒ i < (j + k)
```

FIGURE 3.9. A trait containing a conjecture

problem, but also the conjectures previously proved using the old axioms. LP has facilities that support such regression testing.

LP will, upon request, record a session in a script file that can be replayed. LP "prettyprints" script files, using indentation to reflect the structure of proofs. It also annotates script files with information that indicates when subgoals are introduced (e.g., in a proof by induction), and when subgoals and theorems are proved. On request, as LP replays a script file, it will halt replay at the first point where the annotations and the new proof diverge. This checking makes it easier to keep proof attempts from getting "out of sync" with their author's conception of their structure.

A SMALL PROOF

Figure 3.9 contains a short LSL specification, including a simple conjecture (following the reserved word implies) that is supposed to follow from the axioms. Figure 3.10 shows a script for an LP proof of that conjecture.

The declare commands introduce the variables and operators in the LSL specification. The assert commands supply the LSL axioms relating the operators; the Nat generated by assertion provides an induction scheme for Nat. The prove command initiates a proof by

```
set name nat
declare sort Nat
declare variables i, j, k: Nat
declare operators
    0: → Nat
    s: Nat → Nat
    +: Nat, Nat → Nat
    <: Nat, Nat → Bool
    ..
assert Nat generated by 0, s
assert ac +
assert
    i + 0 == i
    i + s(j) == s(i + j)
    ¬(i < 0)
    0 < s(i)
    s(i) < s(j) == i < j
    ..
set name lemma
prove i < j ⇒ i < (j + k) by induction on j
    <> 2 subgoals for proof by induction on j
    [] basis subgoal
    resume by induction on i
    <> 2 subgoals for proof by induction on i
        [] basis subgoal
        [] induction subgoal
    [] induction subgoal
    [] conjecture
qed
```

FIGURE 3.10. Sample LP proof script

induction of the conjecture. The *diamond* (<>) annotations are provided by LP; they indicate the introduction of subgoals for the inductions. The *box* ([]) annotations are also provided by LP; they indicate the discharge of subgoals and, finally, of the main proof. The `resume` command starts a nested induction. No other user intervention is needed to complete this proof. The `qed` command on the last line asks LP to confirm that there are no outstanding conjectures.

3.6   Lexical and typographic conventions

The Larch languages were designed for use with an open-ended collection of programming languages, support tools, and input/output facilities, each of which may have its own lexical conventions and capabilities. To avoid conflicts, LSL assigns fixed meanings to only a small number of characters. To conform to local conventions and to exploit locally available capabilities, LSL's character and token classes are extensible, and can be tailored for particular purposes by *initialization files*. Since LSL terms appear in interface specifications, corresponding extensibility is a part of each interface language. Appendix C explains the structure of these files and gives the initialization files used in checking the specifications in this book.

There are several semantically equivalent forms of each Larch language. Any of these forms can be translated mechanically into any other without losing information.

- *Presentation forms* are used in environments, such as this book, that have rich character sets with symbols such as $\forall$, $\exists$, $\wedge$, $\vee$, $\in$.

- *Interchange form* is an encoding of the language using a widely available subset of the ISO Latin[4] character set. Characters outside this subset are represented by *extended characters*—sequences of characters from the subset, preceded by a backslash (or other designated character). Interchange form is the "lowest common denominator" for each Larch language. Each Larch tool can parse it and generate it on demand.

- *Interactive forms* may be used by Larch editors, browsers, checkers, etc., for interaction with users. Many such forms will not be limited

---

[4]This is also a subset of the older ASCII subset.

to character strings for input and output (e.g., they will use menus and pointing), and some may impose additional constraints and equivalences (e.g., case folding, operator precedence).

Chapter 4

LSL: The Larch Shared Language

This chapter provides a tutorial introduction the the Larch Shared Language (LSL). It begins by systematically working through the features of the language, illustrating each with a short example. It concludes with a slightly longer example, designed to illustrate how the various features of the language can be used in concert.

4.1 Equational specifications

LSL's basic unit of specification is a *trait*. Consider, for example, the specification of tables that store values in indexed places, Figure 4.1. This is similar to a conventional algebraic specification, as it would be written in many languages [4, 20, 24, 96].

The trait can be referred to by its name, Table1. This should not be

```
Table1: trait
introduces
  new: → Tab
  add: Tab, Ind, Val → Tab
  __ ∈ __: Ind, Tab → Bool
  lookup: Tab, Ind → Val
  isEmpty: Tab → Bool
  size: Tab → Int
  0,1: → Int
  __ + __: Int, Int → Int
asserts ∀ i, i1: Ind, val: Val, t: Tab
  ¬(i ∈ new);
  i ∈ add(t, i1, val) == i = i1 ∨ i ∈ t;
  lookup(add(t, i, val), i1) ==
    if i = i1 then val else lookup(t, i1);
  size(new) == 0;
  size(add(t, i, val)) ==
    if i ∈ t then size(t) else size(t) + 1;
  isEmpty(t) == size(t) = 0
```

FIGURE 4.1. A table trait

confused with the name of a data abstraction (e.g., the sort Tab) or operator (e.g., lookup). The name of a trait is independent of the names that appear within it.

The part of the trait following introduces declares a list of *operators*, each with its *signature* (the *sorts* of its *domain* and *range*). As discussed in Chapter 2, an operator stands for a total function that maps a tuple of values of its domain sorts to its range sort. Every operator used in a trait must be declared; signatures are used to sort-check *terms* in much the same way as expressions are type-checked in programming languages. Sorts are denoted by identifiers and are declared implicitly by their appearance in signatures.

The remainder of this trait constrains the operators by means of *equations*. An equation consists of two terms of the same sort, separated by = or ==. The operators = and == are semantically equivalent, but have a different precedence, as discussed below. We use == as the main connective in equations. Equations of the form *term* == true can be abbreviated by simply writing *term*; thus the first equation in Table1 is an abbreviation for

```
¬ (i ∈ new) == true
```

Double underscores (__) in an operator declaration indicate that the operator will be used in *mixfix terms*. For example, ∈ is declared as a binary infix operator. Infix, prefix, postfix, and distributed operators (such as __+__, -__, __!, {__}, __[__], and if__then__else__) are integral parts of many familiar mathematical and programming notations, and their use can contribute substantially to the readability of specifications.

LSL's grammar for mixfix terms is intended to ensure that legal terms parse as readers expect—even without studying the grammar.[1] LSL has a simple precedence scheme for operators:

- postfix operators that consist of a dot followed by an identifier (as in field selectors, e.g., .first) bind most tightly;

- other user-defined operators and the built-in Boolean negation operator ¬ bind more tightly than

- the built-in equality operators (= and ≠), which bind more tightly than

---

[1]However, writers of specifications should take pity on readers and study the grammar.

- the built-in propositional connectives (∧, ∨, and ⇒), which bind more tightly than

- the built-in conditional connective (if__then__else__), which binds more tightly than

- the equation connective (==).

For example, the equation v == x + w.a.b = y ∨ z is equivalent to the term v = (((x + ((w.a).b)) = y) ∨ z). LSL allows unparenthesized infix terms with multiple occurrences of an operator at the same precedence level, but not different operators; it associates such terms from left to right. Fully parenthesized terms are always acceptable. Thus x ∧ y ∧ z is equivalent to (x ∧ y) ∧ z, but x ∨ y ∧ z must be written as (x ∨ y) ∧ z or as x ∨ (y ∧ z), depending on which is meant.

Each well-formed trait defines a *theory* (a set of sentences closed under logical consequence, see Chapter 2) in multisorted first-order logic with equality. Each theory contains the trait's assertions, the conventional axioms of first-order logic, everything that follows from them, and nothing else. This *loose* semantic interpretation guarantees that formulas in the theory follow only from the presence of assertions in the trait—never from their absence. This is in contrast to algebraic specification languages based on initial algebras [34] or final algebras [85]. Using the loose interpretation ensures that all theorems proved about an incomplete specification remain valid when it is extended.

Each trait should be *consistent:* it must not define a theory containing the equation true == false. Consistency is often difficult to prove and is undecidable in general. Inconsistency is often easier to detect and can be a useful indication that there is something wrong with a trait. Detecting inconsistencies is discussed in Chapter 7.

4.2   Stronger theories

Equational theories are useful, but a stronger theory is often needed, for example, when specifying an abstract type. The constructs generated by and partitioned by provide two ways of strengthening equational specifications.

A *generated by* clause asserts that a list of operators is a complete set of *generators* for a sort. That is, each value of the sort is equal to one that can be written as a finite number of applications of just those operators, and variables of other sorts. This justifies a *generator induction schema* for proving things about the sort. For example, the natural numbers are generated by 0 and succ, and the integers are generated by 0, succ, and pred.

The assertion

```
Tab generated by new, add
```

if added to Table1, could be used to prove theorems by induction over new and add, since, according to this assertion, any value of sort Tab can be constructed from new by a finite number of applications of add. For example, to prove

```
∀ t:Tab, i:Ind (i ∈ t ⇒ size(t) > 0)
```

one can do an inductive proof with the structure

- Basis step:

```
∀ i:Ind (i ∈ new ⇒ size(new) > 0)
```

- Induction step:

```
∀ t:Tab, i1:ind, v1:Val
   (∀ i:Ind (i ∈ t ⇒ size(t) > 0)
     ⇒ (∀ i:Ind (i ∈ add(t, i1, v1)
          ⇒ size(add(t, i1, v1)) > 0)))
```

A *partitioned by* clause asserts that a list of operators constitutes a complete set of *observers* for a sort. That is, all distinct values of the sort can be distinguished using just those operators. Terms that are not distinguishable using any of them are therefore equal. For example, sets are partitioned by ∈, because sets that contain the same elements are equal. Each partitioned by clause is a new axiom that justifies a deduction rule for proofs about values of the sort. For example, the assertion

```
Tab partitioned by ∈, lookup
```

adds the deduction rule

```
∀ i1:ind (i1 ∈ t1 = i1 ∈ t2),
∀ i1:ind (lookup(t1, i1) = lookup(t2, i1)))
---------------------------------------
              t1 = t2
```

If added to `Table1` this partitioned by clause could be used to derive theorems that do not follow from the equations alone. For example, to prove the commutativity of `add` of the same value,

```
∀ t:Tab, i,i1:Ind, v:Val
   (add(add(t, i, v), i1, v)
      = add(add(t, i1, v), i, v))
```

one discharges the two subgoals

```
∀ i2:Ind
   (i2 ∈ add(add(t, i, v), i1, v)
      = i2 ∈ add(add(t, i1, v), i, v))
∀ i2:Ind
   (lookup(add(add(t, i, v), i1, v), i2)
      = lookup(add(add(t, i1, v), i, v), i2))
```

4.3   Combining traits

`Table1` contains three operators that it does not define: `0`, `1`, and `+`. Without more information about these operators, the definition of `size` is not particularly useful. We could add assertions to `Table1` to define these operators. However, it is often better to specify such operators in a separate trait that is included by reference. This makes the specification more structured and makes it easier to reuse existing specifications, such as the traits given in Appendix A. We might remove the explicit introductions of these operators in `Table1`, and instead add an *external reference* to the trait `Integer` (page 163):

```
includes Integer
```

which not only introduces the operators, but also defines their properties.

The theory associated with an including trait is the theory associated with the union of its `introduces` and `asserts` clauses with those of its included traits.

It is often convenient to combine several traits dealing with different aspects of the same operator. This is common when specifying something that is not easily thought of as a data type. Consider, for example, the specifications of properties of relations in Figure 4.2. The trait `equivalence1` has the same associated theory as the less structured trait `equivalence2`.

```
reflexive: trait
   introduces __ ◇ __: T, T → Bool
   asserts ∀ x: T
     x ◇ x

symmetric: trait
  introduces __◇ __: T, T → Bool
   asserts ∀ x, y: T
     x ◇ y == y ◇ x

transitive: trait
   introduces __ ◇ __: T, T → Bool
   asserts ∀ x, y, z: T
     (x ◇ y ∧ y ◇ z) ⇒ x ◇ z

equivalence1: trait
   includes reflexive, symmetric, transitive

equivalence2: trait
   introduces __ ◇ __: T, T → Bool
   asserts ∀ x, y, z: T
     x ◇ x;
     x ◇ y == y ◇ x;
     (x ◇ y ∧ y ◇ z) ⇒ x ◇ z
```

FIGURE 4.2. Specifications of kinds of relations

```
equivalence: trait
  includes
    (reflexive, symmetric, transitive)(≡ for ◇)
```

FIGURE 4.3. An example of renaming

4.4 Renaming

The trait `equivalence1` relies heavily on the use of the same operator symbol, ◇, and the same sort identifier, T, in the three included traits. In the absence of such happy coincidences, renaming can be used to make names coincide, to keep them from coinciding, or simply to replace them with more suitable names, as in Figure 4.3, where ◇ is replaced by a more customary symbol for an equivalence relation.

In general, the phrase *Tr* (*name1* `for` *name2*) stands for the trait *Tr* with every occurrence of *name2* (which must be either a sort or an operator) replaced by *name1*. If *name2* is a sort, this renaming changes the signatures of all of the operators in *Tr* in whose signatures *name2* appears.

The two specifications in Figure 4.4 have the same theory. Note that the infix operator `__∈__` was replaced by the operator `defined`, and that the operator `lookup` was replaced by the mixfix operator `__[__]`. All renamings preserve the order of operands.

Any sort or operator in a trait can be renamed when that trait is referenced in another trait. Some, however, are more likely to be renamed than others. It is often convenient to single these out so that they can be renamed positionally. For example, if the header for the trait had been

```
    SparseArray(Val, Arr): trait
```

the reference

```
    includes SparseArray(Int, IntArr)
```

would be equivalent to

```
    includes SparseArray(Int for Val, IntArr for Arr)
```

4.5 Stating intended consequences

It is not possible to prove the "correctness" of a specification, because there is no absolute standard against which to judge correctness. But since

```
SparseArray: trait
    includes Table1(Arr for Tab, defined for ∈,
        assign for add, __[__] for lookup, Int for Ind)

SparseArrayExpanded: trait
    introduces
        new: → Arr
        assign: Arr, Int, Val → Arr
        defined: Int, Arr → Bool
        __[__]: Arr, Int → Val
        isEmpty: Arr → Bool
        size: Arr → Int
        0,1: → Int
        __ + __: Int, Int → Int
asserts ∀ i, i1: Int, val: Val, t: Arr
    ¬defined(i, new);
    defined(i, assign(t, i1, val)) ==
                        i = i1 ∨ defined(i, t);
    assign(t, i, val)[i1] ==
                        if i = i1 then val else t[i1];
    size(new) == 0;
    size(assign(t, i, val)) ==
        if defined(i, t) then size(t) else size(t) + 1;
    isEmpty(t) == size(t) = 0
```

FIGURE 4.4. Two specifications of sparse arrays

specifications can contain errors, specifiers need help in locating them. LSL specifications cannot, in general, be executed, so they cannot be tested in the way that programs are commonly tested. LSL sacrifices executability in favor of brevity, clarity, flexibility, generality, and abstraction. To compensate, it provides other ways to check specifications.

This section briefly describes ways in which specifications can be augmented with redundant information to be checked during validation. Chapter 7 discusses the use of LP, the Larch proof assistant, in specification debugging.

Checkable properties of LSL specifications fall into three categories: *consistency*, *theory containment*, and *completeness*. As discussed earlier, the requirement of consistency means that any trait whose theory contains the equation `true == false` is illegal.

Implies clauses make claims about theory containment. Suppose we think that a consequence of the assertions of `SparseArray` is that no array with a defined element is empty. To formalize this claim, we could add to `SparseArray`

```
implies ∀ a: Arr, i: Int
    defined(i, a) ⇒ ¬isEmpty(a)
```

The theory to be implied can be specified using the full power of LSL, including equations, generator clauses, partitioning clauses, and references to other traits. Attempting to verify that such a theory actually is implied can be helpful in error detection, as discussed in Chapter 7. Implications also help readers confirm their understanding. Finally, they can provide useful lemmas that will simplify reasoning about specifications that use the trait.

LSL does not require that each trait define a *complete theory*, that is, one in which each sentence is either true or false. Many finished specifications (intentionally) do not fully define all their operators. Furthermore, it can be useful to check the completeness of some definitions long before finishing the specification they are part of. Therefore, instead of building in a single test of completeness that is applied to all traits, LSL provides a way to include within a trait specific checkable claims about completeness, using `converts` clauses.

Adding the claim

```
implies converts isEmpty
```

to `Table1` says that the trait's axioms fully define `isEmpty`. This means that, if the interpretations of all the other operators are fixed, there is only one interpretation of `isEmpty` that satisfies the axioms. (A more complete discussion of the meaning of `converts` is contained in Section 7.1.)

The stronger claim

```
implies converts isEmpty, lookup
```

however, cannot be verified, because the meaning of terms of the form `lookup(new, i)` is not defined by the trait. This incompleteness in `Table1` could be resolved by adding another axiom to the trait, perhaps

```
lookup(new, i) == errorVal
```

But it is generally better not to add such axioms. The specifier of `Table1` should not be concerned with whether the sort `Val` has an `errorVal` and should not be required to introduce irrelevant constraints on `lookup`. Extra axioms give readers more details to assimilate; they may preclude useful specializations of a general specification; sometimes there simply is no reasonable axiom that would make an operator convertible (consider division by 0).

LSL provides an `exempting` clause that lists terms that are not claimed to be defined.[2] The claim

```
implies converts isEmpty, lookup
   exempting ∀ i: Ind lookup(new, i)
```

means that `isEmpty` and `lookup` are fully defined by the trait's axioms plus interpretations of the other operators and of all terms of the form `lookup(new, i)`. This is provable from the specification of `Table1`.

4.6   Recording assumptions

Many traits are suitable for use only in certain contexts. Just as we write preconditions that document when a procedure may properly be called, we write *assumptions* in traits that document when a trait may properly be included. As with preconditions, assumptions impose a proof obligation on the client, and may be presumed within the trait containing them.

It is useful to construct general specifications that can be specialized in a variety of ways. Consider, for example, the specification in Figure 4.5. We might specialize this to `IntegerBag` by renaming `E` to `Int` and

---

[2]This is different from "that are claimed not to be defined."

```
Bag0(E): trait
  introduces
    { }: → B
    insert, delete: E, B → B
    __ ∈ __: E, B → Bool
  asserts
    B generated by { }, insert
    B partitioned by delete, ∈
    ∀ b: B, e, e1, e2: E
      delete(e, { }) == { };
      delete(e1, insert(e2, b)) ==
        if e1 = e2 then b
        else insert(e2, delete(e1, b));
      ¬(e ∈ { });
      e1 ∈ insert(e2, b) == e1 = e2 ∨ e1 ∈ b
```

FIGURE 4.5. A specification of bags

```
Bag1(E): trait
  includes Bag0, Integer
  introduces
    rangeCount: E, E, B → Int
    __ < __: E, E → Bool
  asserts ∀ e1, e2, e3: E, b: B
    rangeCount(e1, e2, { }) == 0;
    rangeCount(e1, e2, insert(e3, b)) ==
      rangeCount(e1, e2, b)
        + (if e1 < e3 ∧ e3 < e2 then 1 else 0)
```

FIGURE 4.6. A specialization of Bag0

including it in a trait in which operators dealing with Int are specified, for example,

```
IntegerBag: trait
  includes Integer, Bag0(Int)
```

The interactions between Integer and Bag0 are limited. Nothing in Bag0 depends on any particular operators being introduced in including traits, let alone their having any special properties. Therefore Bag0 needs no assumptions.

Consider, however, extending Bag0 to Bag1 by adding an operator, rangeCount, to count the number of entries in a B that lie between two values, as in Figure 4.6.

As written, Bag1 says nothing about the properties of the < operator. But it probably doesn't make sense in any specialization unless < provides an ordering on the values of sort E. We cannot define < within Bag1, because it will depend on the trait using Bag1. What we need is an *assumes clause*, as in Figure 4.7.

Since Bag2 may presume its assumptions, its (local) theory is the same as if TotalOrder(E), page 194, had been included rather than assumed; Bag2 inherits all the introductions and assertions of TotalOrder. Therefore, the assumption of TotalOrder can be used to derive various properties of Bag2, for example, that rangeCount is monotonic in its second argument, as claimed in the implies clause.

The difference between assumes and includes appears when Bag2 is used in another trait. Whenever a trait with assumptions is included or assumed, its assumptions must be *discharged*. For example, in

```
Bag2(E): trait
  assumes TotalOrder(E)
  includes Bag0, Integer
  introduces rangeCount: E, E, B → Int
  asserts ∀ e1, e2, e3: E, b: B
    rangeCount(e1, e2, { }) == 0;
    rangeCount(e1, e2, insert(e3, b)) ==
      rangeCount(e1, e2, b)
        + (if e1 < e3 ∧ e3 < e2 then 1 else 0)
  implies ∀ e1, e2, e3: E, b: B
    e1 ≤ e2 ⇒
      rangeCount(e3, e1, b) ≤ rangeCount(e3, e2, b)
```

FIGURE 4.7. An example of an assumption

```
IntegerBag1: trait
  includes Integer, Bag2(Int)
```

the assumption to be discharged is that the (renamed) theory associated with `TotalOrder` is a subset of the theory associated with the rest of `IntegerBag1` (i.e., `Integer`). When a trait includes a trait with assumptions, it is often possible to confirm that these assumptions are *syntactically discharged* by noticing that the same traits are assumed or included by the including trait. For example, the `Integer` trait, page 163 directly includes `TotalOrder`. A more complete discussion of how assumptions are discharged is contained in Chapter 7.

4.7   Built-in operators and overloading

In our examples, we have freely used the predicate connectives defined in Chapter 2. We have also used some heavily overloaded and apparently unconstrained operators: `if__then__else__`, =, and ≠. These operators are built into the language. This allows them to have appropriate syntactic precedence. More importantly, it guarantees that they have consistent meanings in all LSL specifications, so readers can rely on their intuitions about them.

Similarly, LSL recognizes decimal numbers, such as 0, 24, and 1992, without explicit declarations and definitions. In principle, each literal could be defined within LSL, but such definitions are not likely to advance anyone's understanding of the specification. `DecimalLiteral`,

```
OrderedString(E, Str): trait
  assumes TotalOrder(E)
  includes DerivedOrders(Str)
  introduces
    empty: → Str
    __ -| __ : E, Str → Str
    __ < __ : Str, Str → Bool
  asserts
    Str generated by empty, -|
  ∀ e, e1: E, s, s1: Str
      empty < (e -| s);
      ¬ (s < empty);
      (e -| s) < (e1 -| s1) ==
          e < e1 ∨ (e = e1 ∧ s < s1)
  implies TotalOrder(Str)
```

FIGURE 4.8. An example of overloading

page 164 is a predefined quasi-trait that implicitly defines all the numerals
that appear in a specification.

In addition to the built-in overloaded operators and numbers, LSL
provides for user-defined overloadings. Each operator must be declared
in an introduces clause and consists of an identifier (e.g., empty)
or operator symbol (e.g., __<__) and a signature. The signatures of most
occurrences of overloaded operators are deducible from context. Consider,
for example, Figure 4.8.[3] The operator symbol < is used in the last equation
to denote two different operators, one relating terms of sort Str, and the
other, terms of sort E, but their contexts determine unambiguously which
is which.

LSL provides notations for disambiguating an overloaded operator when
context does not suffice. Any subterm of a term can be qualified by its sort.
For example, a:S in a:S = b explicitly indicates that a is of sort S.
Furthermore, since the two operands of = must have the same sort, this
qualification also implicitly defines the signatures of = and b. The last
axiom in Figure 4.8 could also be written as

```
(e -| s):Str < (e1 -| s1):Str ==
    e:E < e1:E ∨ (e = e1 ∧ s:Str < s1:Str)
```

---

[3]DerivedOrders is in Appendix A, page195. It relates the ordering relations ≤, ≥,
<, and > to each other.

```
introduces
  cold, warm, hot:  → Temp
  succ: Temp  → Temp
asserts
  Temp generated by cold, warm, hot
  equations
    cold ≠ warm;
    cold ≠ hot;
    warm ≠ hot;
    succ(cold) == warm;
    succ(warm) == hot
```

FIGURE 4.9. Expansion of an enumeration shorthand

Outside of terms, overloaded operators can be disambiguated by directly affixing their signatures, for example

```
implies converts <:Str,Str→Bool
```

4.8   Shorthands

Enumerations, tuples, and unions provide compact, readable representations for common kinds of theories. They are syntactic shorthands for things that could be written in LSL without them.

ENUMERATIONS

The enumeration shorthand defines a finite ordered set of distinct constants and an operator that enumerates them. For example,

```
Temp enumeration of cold, warm, hot
```

is equivalent to including a trait with the body appearing in Figure 4.9.

TUPLES

The tuple shorthand is used to introduce fixed-length tuples, similar to records in many programming languages. For example,

```
C tuple of hd: E, tl: S
```

is equivalent to including a trait with the body appearing in Figure 4.10. Each field name (e.g., hd) is incorporated in two distinct operators (e.g., _.hd:C→E and set_hd:C,E→C).

```
introduces
   [__, __]: E, S → C
   __.hd: C → E
   __.tl: C → S
   set_hd: C, E → C
   set_tl: C, S → C
asserts
   C generated by [__, __]
   C partitioned by .hd, .tl
   ∀ e,e1: E, s,s1: S
      ([e, s]).hd == e;
      ([e, s]).tl == s;
      set_hd([e, s], e1) == [e1, s];
      set_tl([e, s], s1) == [e, s1]
```

FIGURE 4.10. Expansion of a tuple shorthand

```
S_tag enumeration of atom, cell
introduces
   atom: A → S
   cell: C → S
   __.atom: S → A
   __.cell: S → C
   tag: S → S_tag
asserts
   S generated by atom, cell
   S partitioned by .atom, .cell, tag
   ∀ a: A, c: C
      atom(a).atom == a;
      cell(c).cell == c;
      tag(atom(a)) == atom;
      tag(cell(c)) == cell
```

FIGURE 4.11. Expansion of a union shorthand

UNIONS

The union shorthand corresponds to the tagged unions found in many programming languages. For example,

```
S union of atom: A, cell: C
```

is equivalent to including a trait with the body appearing in Figure 4.11. Each field name (e.g., atom) is incorporated in three distinct operators (e.g., atom:→S_tag, atom:A→S, and __.atom:S→A).

```
InsertGenerated (E, C): trait
  introduces
    empty: → C
    insert: E, C → C
  asserts
    C generated by empty, insert
```

FIGURE 4.12. InsertGenerated.lsl

4.9 Further examples

We have now covered all the facilities of the Larch Shared Language. The next series of examples illustrates their coordinated use.

The trait InsertGenerated, Figure 4.12, abstracts the common properties of data structures that contain elements, such as sets, bags, queues, stacks, and strings. InsertGenerated is useful both as a starting point for specifications of many different data structures and as an assumption when defining generic operators over such data structures.

The generated by clause in InsertGenerated asserts that each value of sort C can be constructed from empty by repeated applications of insert (i.e., empty and insert constitute a complete set of generators for C). This assertion is carried along when InsertGenerated is included in or assumed by other traits, even if those traits introduce additional operators with range C.

The trait Container, Figure 4.13, includes InsertGenerated. It constrains the operators introduced in InsertGenerated, as well as the operators it introduces. The axioms defining count guarantee that insertions are not lost. This implies, for example, that sets do not satisfy this definition of container. The last axiom asserts that, when applied to a non-empty container, tail removes an element equal to the element returned by head. Notice that these axioms do not imply the stronger property ¬isEmpty(c) ⇒ insert(head(c), tail(c)) = c.

The converts clause adds checkable redundancy to the specification. The implied formula follows from the last axiom and the two axioms defining count. If head were to return something that was not in c, inserting it back in would change the count for that value.

PQueue, Figure 4.14, specializes Container by constraining head and tail in a way that is consistent with the last two axioms of Container. The first implication states a fact that may be helpful in

```
Container (E, C): trait
  includes InsertGenerated, Integer
  introduces
    isEmpty: C → Bool
    count: E, C → Int
    __ ∈ __: E, C → Bool
    head: C → E
    tail: C → C
  asserts
    C partitioned by isEmpty, head, tail
    ∀ e, e1: E, c: C
      isEmpty(empty);
      ¬isEmpty(insert(e, c));
      count(e, empty) == 0;
      count(e, insert(e1, c)) ==
        count(e, c) + (if e = e1 then 1 else 0);
      e ∈ c == count(e, c) > 0;
      ¬isEmpty(c) ⇒
        count(e, insert(head(c), tail(c)))
          = count(e, c)
  implies
    ∀ c: C
      ¬isEmpty(c) ⇒ count(head(c), c) > 0;
    converts isEmpty, count, ∈
```

FIGURE 4.13. Container.lsl

```
PQueue (E, Q): trait
  assumes TotalOrder (E)
  includes Container(Q for C)
  asserts ∀ e, e1: E, q: Q
    head(insert(e, q)) ==
      if isEmpty(q) ∨ e > head(q)
        then e else head(q);
    tail(insert(e, q)) ==
      if isEmpty(q) ∨ e > head(q)
        then q else insert(e, tail(q))
  implies
    ∀ q: Q, e: E
      e ∈ q ⇒ ¬ (e < head(q))
    converts head, tail, isEmpty, count, ∈
      exempting head(empty), tail(empty)
```

FIGURE 4.14. PQueue.lsl

reasoning about PQueue and may help readers solidify their understanding of the trait. The second implication states that the trait fully defines head and tail (except when applied to empty), isEmpty, count, and ∈. The axioms that convert isEmpty, count, and ∈ are inherited from Container.

Unlike the preceding traits in this section, PQueue specifies a complete abstract type constructor. In such a trait there is a distinguished sort, sometimes called the *type of interest* [40] or *data sort*. An abstract type's operators can be categorized as *generators*, *observers*, and *extensions* (sometimes in more than one way). A set of generators produces all the values of the distinguished sort. The extensions are the remaining operators whose range is the distinguished sort. The observers are the operators whose domain includes the distinguished sort and whose range is some other sort. An abstract type specification usually has axioms sufficient to convert the observers and extensions. The distinguished sort is usually partitioned by at least one subset of the observers and extensions.

In the example of PQueue, Q is the distinguished sort, empty and insert form a generator set, tail is an extension, head, isEmpty, count and ∈ are the observers, and head, tail, and isEmpty form a partitioning set.

A good heuristic for writing enough equations to adequately define an abstract type is to write an equation defining the result of applying each observer or extension to each generator. For PQueue, this rule suggests writing equations for

```
 1)  isEmpty(empty)
 2)  count(e, empty)
 3)  e ∈ empty
 4)  head(empty)
 5)  tail(empty)
 6)  isEmpty(insert(e, q))
 7)  count(e, insert(el, q))
 8)  e ∈ insert(el, q)
 9)  head(insert(e, q))
10)  tail(insert(e, q))
```

PQueue contains explicit equations for only the last two of these; it inherits equations for five more from Container. The third and eighth terms in the list do not appear explicitly in equations. Instead, ∈ is defined by relating it directly to count. The remaining two terms, head(empty) and tail(empty), are explicitly exempted.

```
PairwiseExtension (o, ⊙, E, C): trait
  assumes Container(E, C)
  introduces
    __ o __ : E, E → E
    __ ⊙ __ : C, C → C
  asserts ∀ e1, e2: E, c1, c2: C
    empty ⊙ empty == empty;
    (¬isEmpty(c1) ∧ ¬isEmpty(c2))
      ⇒ c1 ⊙ c2 = insert(head(c1) o head(c2),
                          tail(c1) ⊙ tail(c2));
  implies
    converts ⊙
    exempting ∀ e: E, c: C
      empty ⊙ insert(e, c),
      insert(e, c) ⊙ empty

PairwiseSum(C): trait
  assumes Container(Int, C)
  includes Integer, PairwiseExtension(+, ⊕, Int, C)
  implies Associative(⊕, C),
          Commutative(⊕ for o, C for T, C for Range)
```

FIGURE 4.15. Specification of generic operators

The traits `PairwiseExtension` and `PairwiseSum`, Figure 4.15, specify generic operators that can be used with various kinds of containers.

`PairwiseExtension` is a generic trait that may be instantiated using a variety of data structures and operators. Given a container sort and a binary operator, o, on elements, it defines a new binary operator, ⊙, on containers. The result of applying ⊙ to a pair of containers is a container whose elements are the results of applying o to corresponding pairs of their elements. The `exempting` clause indicates that, although the result of applying ⊙ to containers of unequal size is not specified, this is not an oversight.

The trait `PairwiseSum` specializes `PairwiseExtension` by binding o to an operator, +, whose definition is to be taken from the trait `Integer` (page 163). The validity of the implications that ⊕ is associative and commutative stems from the replacement of o by +, whose axioms in the trait `Integer` imply its associativity and commutativity. These implications can be proved by induction over `empty` and `insert`.

Chapter 5

LCL: A Larch Interface Language for C

LCL is a Larch interface language for Standard C. LCL is not a C dialect. Programs specified and developed with LCL are C programs, accepted by ordinary C compilers. Use of LCL will tend to encourage some styles of development, but it does not change the programming language.

This chapter is intended to serve three purposes:

- Present almost all of LCL in enough detail to permit interested readers to start writing their own specifications. If you are interested in doing this, we strongly urge you to get the LCL tools first. The tools are available at no cost, as described in Appendix D.

- Provide examples of how two-tiered specifications are used in practice, not just for C, but for any implementation language. While the syntax for incorporating traits may differ, all Larch interface specifications build upon LSL specifications in approximately the same way.

- Illustrate a style of C programming in which abstract types play a major role. While LCL can be used to specify interfaces in which all types are exposed, that is not the style of programming for which it is best suited. It is certainly not one we would wish to encourage.

Before presenting any interface specifications, we discuss the intended relation between LCL specifications and C programs, how names appearing in LCL specifications are related to values in C computations, and the overall structure of LCL function specifications.

This chapter is intended for C programmers—practicing or potential. We assume some familiarity with C. Readers unfamiliar with C may wish to consult one of the numerous books on C.

5.1 The relation between LCL and C

C is a general and flexible language that is used in many different ways. A common style for organizing a program is as a set of program units, often called *modules*. A module consists of an *interface* and an *implementation*. The interface is a collection of types, functions, variables, and constants for use in other modules, called its *clients*.

A C module *M* is typically represented by three files:

- *M*.c contains most of its implementation, including function definitions and private data declarations.

- *M*.h contains a description of its interface, plus parts of its implementation. Comments provide an informal specification of the module. Type declarations, function prototypes, constant definitions, declarations of external variables, and macro definitions provide all the information about *M* that is needed to compile its clients.

- *M*.o contains its compiled form. Such files are linked together to create executable files.

C modules specified using LCL have two additional files:

- *M*.lcl contains its LCL interface specification—a formal description of the types, functions, variables, and constants provided for clients—together with comments providing informal documentation. It replaces *M*.h as documentation for client programmers. The extra information it provides will also be exploited by a planned *LCLint* tool to perform more extensive checking than an ordinary C lint.

- *M*.lh is a header file derived automatically from *M*.lcl to be included in *M*.h. Mechanical generation of .lh files saves the user from having to repeat information in the .h file, saving work and avoiding an opportunity for error. The implementation portion of *M*.h must still be provided by the implementor.

M.lcl may also refer to another kind of file:

- .lsl files contain auxiliary specifications in the form of LSL *traits* to precisely define operators used in .lcl files.

THE LCL STORAGE MODEL

The LCL and LSL tiers of a specification are connected as described in Chapter 3.

Since LCL, like C, is statically typed, the kind of values that an object[1] can be bound to in a state is fixed. Similarly, each LSL value has a unique *sort*. To connect the two languages, there is a mapping from LCL types to LSL sorts. Each built-in type of C, each type built from C type constructors (e.g., int *), and each abstract type defined in LCL is *based on* an LSL sort. If an expression, *e*, denotes an object of type *T* and *T* is based on sort *S*, then the values of *e*^ and *e′* are of sort *S*. The sort on which a type is based does not appear explicitly in LCL specifications. Instead, an LCL type specifier (a type name or an expression denoting a type) is used to stand for its associated sort.

A standard LSL trait defines operators of the sorts upon which C built-in types, e.g., int and double, are based. Users familiar with C will already know what these operators mean. Specifier-supplied traits are used to introduce application-specific operators. Users familiar with the operators involved may not need to examine such traits closely, but most users are expected to read them. A *uses* clause is used to incorporate specifier-supplied traits and to make the connection between types and sorts.

Consider, for example, the specification fragment:

```
uses Vector(int for elem, int[] for vec);

void vMult(int i, int a[]) {
   modifies a;
   ensures a′ = i * a^;
   }
```

The *uses clause* incorporates the trait Vector (not shown here) with two renamings, the sort of the values contained by objects of type int for elem and the sort of the values contained by objects of type int[] for vec. The operator * used in the ensures clause is defined in that trait. The equation containing this operator sort checks because the formal i denotes an int, the formal a an array object, and the expressions a^ and a′ vectors of integers.

VARIABLES, TYPES, OBJECTS AND STATES

Associated with each scope in a C program is an *environment* that maps *variables* to typed objects. A *type*, as we said in Chapter 1, is most

---

[1]Unfortunately, "object" means several different things in different programming languages. In this chapter, we use it in its C sense: memory locations that can contain values; in the next chapter, in its Modula-3 sense.

conveniently thought of as a set of values and a set of operations that can be applied to those values.

LCL provides two different kinds of types. *Exposed types* correspond exactly to types in C; *abstract types*[2] do not correspond to anything in C.

Although C provides no direct support for abstract types, there is a style of C programming in which they play a prominent role. The programmer relies on conventions to ensure that the implementation of an abstract type can be changed without affecting the correctness of clients. The key restriction is that clients never directly access the representation of an abstract value. All access is through the functions provided in its interface. LCL supports this style of programming by providing both *mutable* and *immutable* abstract types. Values of immutable types are used in much the same way as values of exposed types. Values of mutable types are used to support a more object-oriented style of programming.

In LCL, type checking for exposed types follows the usual C rules. For abstract types, however, type checking is done strictly on the basis of names.

Abstractly, an *object* is a container for values of a particular type. More concretely, it can be thought of as region of storage. The major kinds of values are:

- *basic values*. These are mathematical abstractions, like the integer 3, the letter A, and the set $\{3\}$. Such values are independent of the state of any computation. LSL traits are used to give meaning to basic values.

- *structs*. These are (possibly heterogeneous) collections of objects, each denoted by a *field name*. For example, given the variable declaration

  ```
  struct {int fieldA; char fieldB;} s;
  ```

  s.fieldA denotes an object of type intObj and s.fieldB an object of type charObj.

- *unions*. These are somewhat similar to the variant records of other programming languages. They are like structs, except that their objects overlap in memory.

---

[2]See Chapter 3 for a more thorough discussion of abstract types.

- *pointers*. These point to an object in a homogeneous sequence of objects. If p is a pointer, `* (p + minIndex (p))` denotes the first object of the sequence and `* (p + maxIndex (p))` denotes the last object, where `minIndex (p) ` $\leq$ `0` $\leq$ `maxIndex (p)`.[3]

  The infix operator `->` is a syntactic shorthand to dereference a pointer to a struct and then select one of its members. For example, `a->b` is equivalent to `(*a) .b`.

- *arrays*. These are homogeneous sequences of objects. If a is an array, `a [0]` denotes the first object in the sequence and `a [maxIndex (a)]` the last object.

  Although C makes little distinction between pointer and array parameters, LCL treats them rather differently. In a C function prototype, for example, `char  *s` and `char  s []` are equivalent. In an LCL prototype, however, `char  *s` allows access to all of the characters from `* (s + minIndex (s))` (recall that `minIndex` is non-positive) to `* (s + maxIndex (s))`, while `char  s []` allows access only to the characters from `s [0]` to `s [maxIndex (s)]`.

- *objects*. The value of an object may itself be an object. For example, the value of a field of a struct may itself be a struct. The value of an object of a mutable abstract type is always an object. Therefore, if x is a formal parameter of a mutable abstract type, x^ stands for the value contained in the pre-state by the object to which x is bound.

A function call may change the values of objects accessible to the caller, but it cannot change the caller's environment. Therefore, for our purposes, the *state* of a C computation can be thought of as a mapping from objects to values.

Since parameters are passed by value in C, formal parameters should be thought of as denoting values.[4] In the case of formals that are of type array or of a mutable abstract type, this value is an object. Global variables always denote objects.

---

[3]C does not make the values of `maxIndex` and `minIndex` available at runtime, but they are useful for specifying and reasoning about programs.

[4]Within the body of a function, an object is associated with a formal parameter, but since that object does not exist in the environment of the caller, it is not relevant to the specification.

In LCL, the postfix operators ^ and ′ are used to refer to the values contained by objects in the *pre-state* and *post-state* of a function. They can be applied to expressions denoting objects, collections of objects, or sequences of objects.

- When applied to an object, ^ and ′ yield the value stored in that object. For example, if x is a global variable of type int, x^ = 3 asserts that in the pre-state the value contained by the object to which the variable x is bound is 3. On the other hand, if x is a formal parameter of type int, x^ = 3 does not sort check, since ^ cannot be applied to a basic value.

- When applied to an array, ^ and ′ yield a vector containing the values stored in the sequence of objects denoted by the array.

- When applied to a struct, ^ and ′ yield a tuple containing the values stored in the collection of objects denoted by the struct. Here again, we make a distinction between pointers and arrays. If a field of the struct has an array type, the tuple contains a vector. If the field has a pointer type, the tuple contains a pointer.

5.2   Function specifications

A C function may communicate with its callers by returning a result, by accessing objects accessible to the caller, by modifying such objects, or by returning control to a different place. The specification of each function in an interface can be studied, understood, and used without reference to the specifications of other functions. As discussed in Chapter 3, a specification consists of a *function header* (similar to a C function prototype) followed by a body. Recall that the specification places constraints on both clients and implementations of the function.

- The *requires clause* (precondition) restricts the state and arguments with which the client is allowed to call the function; the implementor may presume it on entry. An omitted requires clause is equivalent to the weakest possible requirement, requires true.

- The *modifies clause* says what a function is allowed to change. If there is no modifies clause, then nothing may be changed.

- The *ensures clause* (postcondition) places constraints on the function's behavior when it is called properly. It relates the state when the function is called, the *pre-state,* and the state when it terminates, the *post-state.* The object result contains the value (if any) returned by the function, and the object control contains the point to which control will be transferred.[5]

- The client is expected to establish the precondition before each call; having done so, the client may presume that the function will terminate in a state satisfying the postcondition, with changes limited to the modifies list.

- The implementor may presume the precondition upon each entry. Under that presumption, the implementation must terminate in a state satisfying the postcondition, without changing any client-visible object not in the modifies list.

5.3   A guided tour through an LCL specification

To illustrate the use of most of LCL's features, we present and discuss a small specification. This example is only superficially realistic; it was structured to use language constructs in the order we want to discuss them. It is not really a typical specification or an especially wonderful program design. As you study this tutorial, you will probably find it instructive to consider alternative designs and how they would be specified.

The example in this section uses various conventions for names, formatting, comments, etc. These are not mandated by LCL; specifications should be written using the conventions of the organization for which they are intended. Because the example is being used to document LCL features, rather than a real interface, the density of comments embedded within the formal text is low, and most of the comments are in the accompanying prose.

This example has been machine-checked for syntax and type correctness. The .lcl and .lsl files have been checked by the LCL and LSL Checkers, respectively. The .lh files were automatically generated by the LCL Checker. The .lh, .h, and .c files were compiled by gcc (this

---

[5]In the pre-state, the value of control is the return address of the invocation. Constructs like abort and longjmp can be specified as modifications of control.

took longer than all the Larch checking). Finally, the compiled code was exercised by a test driver. Although we tried to be careful at each stage of development, each of the mechanical checks caught errors that we had not.

The example is a very simplified employee data base. We

- start with a couple of traits defining useful operators on strings,

- move to a simple interface using exposed types to represent individual employee records,

- introduce an abstract data type for representing sets of employees,

- specify the database interface,

- present a small test program,

- specify some modules that will be used in the implementation, and

- comment on the implementations.

STRING TRAITS

The traits in Figures 5.1 and 5.2 present a collection of operators on strings. They are used throughout the interface specifications in this section.

The trait cstring, Figure 5.1, specializes the strings of the String trait in Appendix A (page 173) to the null-terminated strings conventionally used in C programs. Note that this trait defines the operators throughNull, sameStr and lenStr only when they are applied to null-terminated strings.

The trait sprint was written for specifying functions that convert values to strings. It is intentionally weak. It doesn't say much about the meanings of its operators. This allows considerable flexibility in implementing the interface functions. The first equation guarantees that different T values will have different string forms, without specifying what those forms are. The second equation gives two important properties of acceptable string forms. We could repeat these properties in the interface specification of each such function, but it is better to get them right once, and then reuse the trait.

EMPLOYEE

The interface specified in Figure 5.3, employee, exports two constants, three types, and four functions to its clients.

```
cstring: trait
  includes String(char, String), Integer(int for Int)
  introduces
    null: → char
    nullTerminated: String → bool
    throughNull: String → String
    sameStr: String, String → bool
    lenStr: String → int
  asserts
    ∀ s, s1, s2: String, c: char
      ¬nullTerminated(empty);
      nullTerminated(s ⊢ c) ==
        c = null ∨ nullTerminated(s);
      nullTerminated(s)
        ⇒ throughNull(s ⊢ c) = throughNull(s);
      ¬nullTerminated(s)
        ⇒ throughNull(s ⊢ null) = s ⊢ null;
      sameStr(s1, s2) ==
          throughNull(s1) = throughNull(s2);
      lenStr(s) == len(throughNull(s)) - 1
```

FIGURE 5.1. cstring.lsl fragment

```
sprint(T, String): trait
  includes cstring
  introduces
    parse: String → T
    unparse: T → String
    isSprint: String, T → bool
  asserts ∀ t: T, s: String
    parse(unparse(t)) == t;
    isSprint(s, t) ==
        parse(s) = t ∧ nullTerminated(s)
  implies T partitioned by unparse
```

FIGURE 5.2. sprint.lsl

```
constant int maxEmployeeName;
constant int employeePrintSize;

typedef enum {MALE, FEMALE, gender_ANY} gender;
typedef enum {MGR, NONMGR, job_ANY} job;
typedef struct {int ssNum;
                char name[maxEmployeeName];
                int salary;
                gender gen;
                job j;} employee;

uses employeeConstants, sprint(employee, char[]);

void employee_sprint(char s[], employee e) {
    requires maxIndex(s) ≥ employeePrintSize;
    modifies s;
    ensures isSprint(s', e)
      ∧ lenStr(s') = employeePrintSize;
    }
bool employee_equal(employee *e1, employee *e2) {
    ensures result = sameStr(e1→name^, e2→name^)
            ∧ (e1→ssNum^ = e2→ssNum^)
            ∧ (e1→salary^ = e2→salary^)
            ∧ (e1→gen^ = e2→gen^)
            ∧ (e1→j^ = e2→j^);
    }
bool employee_setName(employee *e, char na[]) {
    requires nullTerminated(na^);
    modifies e->name;
    ensures result = lenStr(na^) < maxEmployeeName
            ∧ (if result
                then sameStr(e->name', na^)
                    ∧ nullTerminated(e->name')
                else e->name' = e->name^);
    }
void employee_initMod(void) {
    ensures true;
    }
```

FIGURE 5.3. employee.lcl

```
employeeConstants: trait
  assumes CTrait
  introduces
    maxEmployeeName, employeePrintSize: → int
  asserts equations
    maxEmployeeName > 0 ∧ maxEmployeeName ≤ 20;
    employeePrintSize > 0 ∧ employeePrintSize ≤ 80
```

FIGURE 5.4. employeeConstants.lsl

The *constant declarations* give symbolic names for values that are used elsewhere in the specification and that may be used by clients of the interface. The allowable values of the two constants are restricted by axioms in the trait `employeeConstants`, Figure 5.4. LCL interface constants may be implemented either by macro definitions or by C `const` variables.

The interface defines, in the lines that look like C typedefs, three exposed types, `gender`, `job` and `employee`. Clients of this interface are being told exactly how these types are represented, and clients may deal with values of these types in any way allowed by Standard C.

The uses clause in Figure 5.3 directly incorporates two LSL specifications. The trait `employeeConstants`, which was written specifically for use in `employee.lcl`, constrains the values of the two exported constants: any `int` from 1 to 20 is allowed for `maxEmployeeName` and any `int` from 1 to 80 is allowed for `employeePrintSize`. The trait `sprint` gives the meaning of operators such as `isSprint` and `nullTerminated` (recall that `sprint` includes `cstring`) that are used later in the specification. Notice that the use of `sprint` involves a renaming. The sort `T` of `sprint.lsl` is to be replaced by the sort on which the type `employee` is based and the sort `String` by the sort on which the type `char[]` is based.

The specification for each function gives both the precondition that is assumed to hold in the pre-state (when the function is called) and the postcondition that is guaranteed to hold in the post-state (upon return). The function `employee_sprint` is typical of a kind found in many interfaces. It converts `employee` values into a string form suitable for printing, and stores this string in an array. Its specification begins with its *function prototype*. LCL prototypes are more restricted than C's; LCL requires that each of the formal parameters be named, so that the body of the specification can refer to any parameter by name. Since all functions in an interface are exported, the keyword `extern` will be added automatically when `employee.lh` is generated.

The body of the specification consists of three clauses.

- The requires clause says that the array s must be big enough to hold the longest string that could ever be returned.

- The modifies clause says that only the contents of the array s can be changed.

- The ensures clause constrains the new value of s.

A good rule of thumb is that each object in the modifies clause should appear in primed form at least once in the ensures clause—unless it is intentionally not being constrained.

Array parameters are passed as pointers in C; s is a pointer to an array. The term s′ denotes the vector of characters contained by the actual parameter corresponding to s upon return from `employee_sprint`. Since struct parameters are copied, e denotes a value of type `employee`, rather than a pointer.

This specification does not say what string will be generated for each `employee` value—only that it will have certain properties. We might want such freedom, for example, in a module that will have different implementations for different countries, languages, or output devices. This specification does not even require an implementation to be deterministic.[6] Although our implementation of `employee` does not exploit this freedom, later interfaces will have implementations that do exploit allowed non-determinism.

The specification of `employee_equal` may strike the reader as surprisingly complicated. The questions arises, why didn't we use one of the following, simpler, ensures clauses?

```
ensures result = ((*e1) = (*e2))

ensures result = ((*e1)^ = (*e2)^)
```

The first of these clauses asserts that `result` is true exactly when e1 and e2 point to the same struct. This is unlikely to be appropriate. The second clause asserts that `result` is true exactly when e1 and e2 point to structs containing the same values. Even this is likely to be too strong,

---

[6]A function is deterministic if its post-state is completely determined by its pre-state.

since it requires that the arrays containing the names be the same beyond the terminating `null` character.

The function `employee_setName` returns a value of type `bool`, the one built-in type of LCL that is missing from C. When LCL specifications are checked, `bool` is treated as a distinct type.

The requires clause in `employee_setName` says that the function should be called only with null-terminated strings. The implementation is entitled to rely on this. Indeed, it must. It is not generally possible to determine at runtime the `maxIndex` of an array. Yet without a guarantee that a string is null-terminated, it is not safe to search for its terminating null character, because the search might run past the end of the allocated storage and generate references to nonexistent memory. Completely defensive programming just isn't possible in C. The implementation of `employee_setName` in `employee.c`, Figure 5.8, relies on this property from its specification. It may crash if `na^` isn't null-terminated.

The modifies clause says that `employee_setName` may change one field of its first argument, `e->name`, but nothing else. Unlike requires and ensures clauses, a modifies clause constrains everything it doesn't mention.

The ensures clause says that `employee_setName` will have one of two outcomes. It will either:

- Make the `name` field of its first argument the same as its second argument (when both are interpreted as strings), make the new value of the `name` field be null-terminated, and return `TRUE`, or

- Change nothing and return `FALSE`.

Furthermore, the first outcome will occur exactly when the new name fits (i.e., `lenStr(na^) < maxEmployeeName`). The use of `result` in several subterms of an ensures clause is a frequent idiom. Since the predicate in the ensures clause is just a logical formula, it makes no semantic difference whether the equation for `result` is written first or last. We are free to choose an order that helps the exposition or emphasizes some particular aspect of the specification.

In this example, we include an `initMod` function as part of every interface. Later we will discuss the way in which we use these functions. The function `employee_initMod` is required by its specification to have no visible effect, since it modifies nothing and returns no value. The absence of a requires clause (equivalent to `requires true`) says that it must always terminate.

```
/*PASS Output from LCL Version 1.7 11-AUG-1992 */
#include "bool.h"
typedef enum {
    MALE,
    FEMALE,
    gender_ANY} gender;

typedef enum {
    MGR,
    NONMGR,
    job_ANY} job;

typedef struct {
    int ssNum;
    char name[maxEmployeeName];
    int salary;
    gender gen;
    job j;
} employee;

extern void employee_sprint (char s[], employee e);

extern bool employee_equal (employee *e1, employee *e2);

extern bool employee_setName (employee *e, char na[]);

extern void employee_initMod (void);
```

FIGURE 5.5. employee.lh

```
#if !defined(EMPLOYEE_H)
#define EMPLOYEE_H

#define maxEmployeeName (20)
#define employeeFormat "%9d  %-20s  %-6s  %-11s  %6d.00"
#define employeePrintSize (63)

#include "employee.lh"

#define employee_initMod()  bool_initMod()
#endif
```

FIGURE 5.6. employee.h

```
#if !defined(BOOL_H)
#define BOOL_H
#define FALSE 0
#define TRUE (! FALSE)
typedef int bool;
#define bool_initMod()
#endif
```

FIGURE 5.7. bool.h

From the specification in employee.lcl the LCL Checker generates the file employee.lh, Figure 5.5. In addition to the appropriate typedefs and function prototypes, it #includes bool.h, Figure 5.7, for the implicitly imported interface bool. This is used in the implementation of employee.h, Figure 5.6, and indirectly, in employee.c, Figure 5.8.

By convention, we start each .h files with a #if that makes sure that its body will only be included once into any module. Both employee.c and all clients of employee will include employee.h. In turn, employee.h includes employee.lh, which provides prototypes. The implementation of the function employee_initMod is also in employee.h.

The file employee.h, Figure 5.6, contains macros defining the constants maxEmployeeName and employeePrintSize. Because of a restriction imposed by C, the definition of maxEmployeeName must precede the inclusion of employee.lh, since it is used in the typedef of employee. The #define cannot be automatically generated because the LCL processor has no way of knowing what value the constant is to have; the specification leaves that design decision to the implementation.

The file employee.h also implements employee_initMod. Our convention is that each module initializes every module it explicitly imports. Thus employee_initMod calls bool_initMod.[7]

In general, M.h contains, in order:

- A test of whether M_H is #defined in the current context. This make sure that, for example, a client of M can safely include it, and other clients can include them both without getting errors caused by repeated type definitions.

---

[7]Since the specification of employee_initMod guarantees that it modifies nothing, calling it multiple times cannot have effects visible to clients.

```
#include <stdio.h>
#include "employee.h"

bool employee_setName(employee *e, char na []) {
    int i;

    for (i = 0; na[i] != '\0'; i++)
      if (i == maxEmployeeName) return FALSE;
    strcpy(e->name, na);
    return TRUE;
}
bool employee_equal(employee * e1, employee * e2) {
    return ((e1->ssNum == e2->ssNum)
            && (e1->salary == e2->salary)
            && (e1->gen == e2->gen)
            && (e1->j == e2->j)
            && (strncmp(e1->name, e2->name,
                maxEmployeeName) == 0));
}
void employee_sprint(char s[], employee e) {
    static char *gender[] ={"male", "female", "?"};
    static char *jobs[] = {"manager", "non-manager", "?"};

    (void) sprintf(s,
                   employeeFormat,
                   e.ssNum,
                   e.name,
                   gender[e.gen],
                   jobs[e.j],
                   e.salary);
}
```

FIGURE 5.8. employee.c

- A definition of M_H.

- Definitions of all constants declared in M.lcl, either as macros or as C const variables.

- Concrete representations (typedefs) for any abstract types declared in M.lcl.

- A #include of M.lh.

- Macros, if any, for inline implementations of functions with prototypes in M.lh.

EMPSET

Empset, in Figure 5.9, is a mutable abstract type. Values of the type are objects that contain sets of employees. As we have seen, exposed types are specified using C typedefs. Abstract types are specified as collections of functions that manipulate values of the type. The representation of these values is hidden within the implementation. Clients can create, modify and examine empsets by calling the functions specified in the interface, but they cannot directly access the representation of empsets.

Type checking for abstract types in both the LCL Checker and LCLint is based on type names, not on their representations. However, within the implementation of the module exporting an abstract type, LCLint treats the abstract type and its representation as the same. This allows the implementation to access the internal structure that is hidden from clients.

The *imports clause* of empset.lcl says that the specification of the empset interface depends on the specification of employee; it gives empset and its clients access to the constants, types and functions exported by employee. It also makes the trait associated with the employee interface a part of the specification of the empset interface. Such specification dependencies should not be confused with implementation dependencies, where one module is used within the implementation of another. Implementation dependencies are typically a superset of the specification dependencies. Clients, however, should not be concerned with implementation dependencies.

The uses clause brings in two traits. The trait sprint is used in exactly the same way as it was in employee. The invocation of the LSL handbook trait Set (page 167) substitutes the sort on which type employee is based for E and the sort on which type empset is based for C.

```
imports employee;
mutable type empset;
uses Set(employee, empset),
     sprint(empset, char[]);

empset empset_create(void) {
   ensures fresh(result) ∧ result' = { };
   }
void empset_final(empset s) {
   modifies s;
   ensures trashed(s);
   }
void empset_clear(empset s) {
   modifies s;
   ensures s' = { };
   }
bool empset_insert(empset s, employee e) {
   modifies s;
   ensures result = ¬ (e ∈ s^) ∧ s' = insert(e, s^);
   }
void empset_insertUnique(empset s, employee e) {
   requires ¬ (e ∈ s^);
   modifies s;
   ensures s' = insert(e, s^);
   }
bool empset_delete(empset s, employee e) {
   modifies s;
   ensures result = e ∈ s^ ∧ s' = delete(e, s^);
   }
empset empset_union(empset s1, empset s2) {
   ensures result' = s1^ ∪ s2^ ∧ fresh(result);
   }
```

FIGURE 5.9. empset.lcl, part 1

```
empset empset_disjointUnion(empset s1, empset s2) {
    requires s1^ ∩ s2^ = { };
    ensures result' = s1^ ∪ s2^ ∧ fresh(result);
    }
void empset_intersect(empset s1, empset s2) {
    modifies s1;
    ensures s1' = s1^ ∩ s2^;
    }
int empset_size(empset s) {
    ensures result = size(s^);
    }
bool empset_member(employee e, empset s) {
    ensures result = e ∈ s^;
    }
bool empset_subset(empset s1, empset s2) {
    ensures result = s1^ ⊆ s2^;
    }
employee empset_choose(empset s) {
    requires s^ ≠ { };
    ensures result ∈ s^;
    }
char *empset_sprint(empset s) {
    ensures isSprint(result[]', s^)
       ∧ fresh(result[]);
    }
void empset_initMod(void) {
    ensures true;
    }
```

FIGURE 5.9. empset.lcl, part 2

```
empset es1, es2;
es1 = empset_create ( );
es2 = es1;
empset_insert(es2, e);
if (empset_size(es1) == 1) printf("Sharing.");
    else printf("No sharing.");
```

FIGURE 5.10. Code to test for sharing

```
empset es1, es2;
es1 = empset_create ( );
es2 = empset_create ( );
if (es1 == es2) printf("Same object.");
    else printf("Different objects.");
```

FIGURE 5.11. Code to test meaning of ==

Clients may write assignment statements involving variables and values of abstract types. Since the value of an object of a mutable abstract type is itself an object, assignments produce sharing. Consider, for example, the code fragment in Figure 5.10.

Because of the semantics associated with mutable abstract types, this program code will print "Sharing." As we shall see shortly, it is the responsibility of the implementor of the type to ensure that assignment has the proper semantics.

Clients may not write code that uses C's == operator to compare values of abstract types. The problem is that for mutable abstract types an expression of the form x == y would return true exactly when x and y denote the same object. For example, the code in Figure 5.11 would print "Different objects." For immutable abstract types, however, the result of a comparison using == would be unpredictable, since the implementation has the freedom to have or not have multiple copies of the same value. We return to this point in Section 5.3.

The first two functions exported by empset.lcl, empset_create and empset_final, are typical of functions found in interfaces exporting abstract types.

The first conjunct in the ensures clause of empset_create says that the function returns a fresh object of type empset. Saying that it is *fresh* means that it is not aliased to any objects visible in the pre-state. The second conjunct says that the value of the returned object is the empty set of employees. This function will typically appear in a statement of the

form es = empset_create(). Since empsets are mutable, calls
to other functions exported by this interface, such as empset_insert,
can then be used to change the value contained by the object.

A client of empset should call empset_final when it is certain that
an empset object will never be referenced again. The clause ensures
trashed(s) says that upon return from empset_final(es) nothing
can be assumed about the value of the object to which es is bound. The
assertion trashed(s) is not equivalent to

```
modifies s
ensures true
```

because referencing a trashed object can even cause the client program to
crash.

A good implementation of empset_final will free storage that is no
longer needed, although this specification does not require it to. Since
a client has no information about how an empset is represented, it
cannot directly free the storage consumed by an empset. For example,
if empset were implemented as a pointer to a pointer to a data structure,
the call free(es) would free only the pointer, not the data structure.

The third function in the interface, empset_clear, is provided for
reinitializing an existing empset. Unlike empset_create, it does not
create a new empset but rather has a side effect on an existing object.

The functions empset_insert and empset_insertUnique both
add an employee to an empset. The chief difference is that the latter
requires that the employee to be added is not already present. This may
make it possible to implement the function more efficiently. However, if
the requirement is violated, the behavior of empset_insertUnique
is totally unconstrained by the specification. The implementation we give
later does not check the requirement. If it is violated the implementation
returns without complaint, but it breaks a representation invariant—thus
leading to unpredictable behavior on subsequent uses of the empset.

The functions empset_union and empset_disjointUnion both
return the union of two empsets. Once again, the requires clause makes
it possible to implement one more efficiently than the other. Notice that
even though s1 and s2 are not modified, the specifications refer to s1^
and s2^. The ^ is needed because s1 and s2 refer to objects. These must
be evaluated in some state to get a value. Here s1 and s2 contain the same
values in the pre- and post-states. By convention, we use ^ rather than '
for objects that are guaranteed to have the same values in both states.

Since both functions are required (by fresh(result)) to return sets that are not aliased to any objects visible in the pre-state, the sets that they return can be modified without affecting the values of other sets. For example, knowing that the result `empset` is fresh allows the client to pass it to `empset_final` without worrying about having an effect on other `empset`s.

One way of ensuring freshness is to allocate new storage. This raises the question of what happens if there is no storage to allocate. In the implementations of these functions (see Appendix B), this is handled by printing a message and terminating the program.[8] But such behavior seems to violate the specification, which says that they should return. We could have augmented the specification to take the possibility of running out of storage into account, but it would have been tedious and not very informative. Almost every function may fail for lack of storage in the stack or heap. So the possibility of exiting the entire program, instead of returning from the function, is implicit in every `ensures` clause. This allows any function to terminate the program. Of course, responsible implementors do not take wanton advantage of this. For some applications it may be important to specify interfaces that preclude running out of storage.

The requires clause of `empset_choose` is necessary to guarantee that the ensures clause is satisfiable. If `s^` is empty, it is not possible to return an `employee` that is a member. If `s^` contains more than one element, the specification allows any member `s^` to be returned. The implementation we present later gains efficiency by being abstractly non-deterministic: A single abstract `empset` value may have many different representations (depending on the order in which its elements were inserted), and the value returned by `empset_choose` is determined by the representation value passed in.

Although the remaining functions are a necessary part of this interface, they don't illustrate any new LCL features. An implementation of the interface is given in Appendix B.

DBASE

The next specification describes a simple data base of employees.

Up to now we have presented modules by first giving an interface specification, then its auxiliary LSL specification, and finally, its implementation. This works well when the reader has good *a priori* intuition

---

[8]For simplicity, our implementation checks inline after each allocation. In practice, it is better to isolate this by calling user-supplied allocation routines.

about the meaning of the abstractions used in the interface specification. When such intuition cannot be relied upon, it is often better to present the auxiliary specification first, as we do here.

Figure 5.12 contains a trait that constrains the kinds of elements a database may contain. Not coincidentally, it corresponds closely to the trait associated with employee.lcl. It is assumed by the dbase trait.

Figure 5.13 starts by including Set. This inclusion tells us that a db is a set of employees. Recall that employee is defined in dbaseAssumptions to be a tuple with five fields. Notice that since a db is merely a set of tuples, no invariant about the elements, e.g., that no

```
dbaseAssumptions: trait
   includes Set(employee for E, empset for C)
   gender enumeration of MALE, FEMALE, gender_ANY
   job enumeration of MGR, NONMGR, job_ANY
   employee tuple of ssNum: int,
                     name: employee_name,
                     salary: int,
                     gen: gender,
                     j: job
```

FIGURE 5.12. dbaseAssumptions.lsl

```
dbase: trait
   assumes dbaseAssumptions
   includes Set(db for C, employee for E, new for { },
               hire for insert)
   db_q tuple of g:gender, j: job, l: int, h: int
   db_status enumeration of db_OK, salERR, genderERR,
                            jobERR, duplERR
introduces
   query: db, db_q → empset
   match: gender, gender → bool
   match: job, job → bool
   fire, promote: db, int → db
   setSal: db, int, int → db
   find: db, int → employee
   employed: db, int → bool
   numEmployees: db → int
```

FIGURE 5.13. dbase.lsl, part 1

```
asserts
  ∀ e: employee, k: int, g, gq: gender,
            j, jq: job, q: db_q, sal: int, d: db
    query(new, q) == { };
    query(hire(e, d), q) ==
      if match(q.g, e.gen) ∧ match(q.j, e.j)
         ∧ q.l ≤ e.salary ∧ e.salary ≤ q.h
        then insert(e, query(d, q)) else query(d, q);
    match(gq, g) == gq = gender_ANY ∨ g = gq;
    match(jq, j) == jq = job_ANY ∨ j = jq;
    fire(new, k) == new;
    fire(hire(e, d), k) ==
      if e.ssNum = k
        then fire(d, k) else hire(e, fire(d, k));
    promote(new, k) == new;
    promote(hire(e, d), k) ==
      if e.ssNum = k
        then hire(set_j(e, MGR), promote(d, k))
        else hire(e, promote(d, k));
    setSal(new, k, sal) == new;
    setSal(hire(e, d), k, sal) ==
      if e.ssNum = k
        then hire(set_salary(e, sal),
                  setSal(d, k, sal))
        else hire(e, setSal(d, k, sal));
    employed(d, k)
      ⇒ (find(d, k).ssNum = k ∧ find(d, k) ∈ d);
    employed(new, k) == false;
    employed(hire(e, d), k) ==
      e.ssNum = k ∨ employed(d, k);
    numEmployees(new) == 0;
    numEmployees(hire(e, d)) == numEmployees(d)
       + (if employed(d, e.ssNum) then 0 else 1);
```

FIGURE 5.13. dbase.lsl, part 2

two `employees` have the same `ssNum`, is implied. This is in contrast to type db, whose specification, Figure 5.14, does imply such an invariant.

In addition to the operators inherited from `Set`, the trait introduces a number of operators that will prove useful in writing `dbase.lcl`. Understanding the meaning of these operators is the key to understanding `dbase.lcl`, Figure 5.14.

The most interesting of these operators is `query`. The first two axioms imply that the value of `query(d, q)` is the set containing all `employees` in the data base that match the `gender` and `job` fields of q and that have salaries between `q.l` and `q.h`.

The `dbase` interface encapsulates a database and a set of functions to query and manipulate it. It exports two exposed types, db_q and db_status, and a number of functions. It also contains our first use of global variables. LCL uses the same scope rules as C. However, LCL extends the function prototype by including a list of the *global variables* referenced by the function. LCLint will check that each global variable accessed by the function body appears in its globals list.

At first glance, it may seem a bit surprising that we have chosen to make db an immutable type. The reason for this is that we don't intend to have formals of type db. Changes to the global variable d will be described as changes to the binding of the variable, not as mutations to the object to which the variable is bound in the pre-state.

As it happens, the global variable in dbase is a *specification variable*. Such variables are declared solely to facilitate writing specifications. Neither the specification variable d nor the specification type db is exported by the interface. Client code cannot refer to either. Furthermore, since they are not exported, specification types and variables need not be implemented. In fact, neither the type db nor the variable d appears in our implementation of this interface.

This example contains our first use of the an LCL *claims* clause. Such clauses play a role analogous to the `implies` clauses of LSL. They assert facts that the specifier believes should be derivable from the rest of the specification. The `claims` clause here asserts that `ssNums` are unique keys for employees. The term d$^\bullet$ is analogous to d$^\wedge$ and d′ ; it means "the value of d in any state visible to clients of this interface." Therefore, this claim is an invariant that must hold in all states visible to clients. As we shall see shortly, such invariants can be verified by data type induction.

The function `hire` is closely related to the operator `hire` of `dbase.lsl`. The difference is that it does some error checking and returns

```
imports employee, empset, stdio;

typedef struct{gender g; job j; int l; int h;} db_q;
typedef enum {db_OK, salERR, genderERR, jobERR,
              duplERR, missERR} db_status;
spec immutable type db;
spec db d;

uses dbase, sprint(ioStream, db);

claims UniqueKeys(employee e1, employee e2) db d; {
   ensures
      (e1 ∈ d• ∧ e2 ∈ d•  ∧ e1.ssNum = e2.ssNum)
       ⇒ (e1 = e2);
   }

db_status hire(employee e) db d; {
   modifies d;
   ensures
      (if result = db_OK
       then d' = hire(e, d^) else d' = d^)
        ∧ result =
           (if e.gen = gender_ANY then genderERR
            else if e.j = job_ANY then jobERR
            else if e.salary < 0 then salERR
            else if employed(d^, e.ssNum) then duplERR
            else db_OK);
   }
void uncheckedHire(employee e) db d; {
   requires e.gen ≠ gender_ANY ∧ e.j ≠ job_ANY
            ∧ e.salary ≥ 0 ∧ ¬employed(d^, e.ssNum);
   modifies d;
   ensures d' = hire(e, d^);
   }
bool fire(int ssNum) db d; {
   modifies d;
   ensures result = employed(d^, ssNum)
     ∧ d' = fire(d^, ssNum);
   }
```

FIGURE 5.14. dbase.lcl, part 1

```
int query(db_q q, empset s) db d; {
   modifies s;
   ensures s' = s^ ∪ query(d^, q)
           ∧ result = size(query(d^, q));
   }
bool promote(int ssNum) db d; {
   modifies d;
   ensures
     result = (employed(d^, ssNum)
              ∧ find(d^, ssNum).j = NONMGR)
      ∧ (if result then d' = promote(d^, ssNum)
         else d' = d^);
   }
db_status setSalary(int ssNum, int sal) db d; {
    modifies d;
    ensures
      result =
        (if employed(d^, ssNum)
           then (if sal < 0 then salERR else db_OK)
           else missERR)
       ∧ (if result = db_OK
             then d' = setSal(d^, ssNum, sal)
             else d' = d^);
   }
void db_print(void) db d; FILE *stdout; {
    modifies *stdout^;
    ensures ∃ s:ioStream (
                (*stdout^)' = write((*stdout^)^, s)
                ∧ isSprint(d^, s));
   }
void db_initMod(void) db d; {
   modifies d;
   ensures d' = new;
   }
```

FIGURE 5.14. dbase.lcl, part 2

a result indicating the outcome of this checking.

The function `uncheckedHire` does no error checking. Of course, if it is called when its requires clause does not hold, it is likely to do something unfortunate that may not be detected for quite some time. Both functions modify the specification variable `d`. Since `d` is a global variable rather than a formal parameter, it can be accessed directly; there is no need to pass in a pointer to it.

The function `query` is also closely related to the LSL operator `query`. But the operator returns an `empset` and the function returns an `int` equal to the number of employees added to `s` as the required side effect of calling `query`. This is a common C idiom.

Now we can use data type induction, discussed in Chapter 3, to show that the claims clause holds. The function `dbase_initMod` ensures that `d` starts out empty. The only functions that are allowed to add employees to `d` are `hire` and `uncheckedHire`. If `hire` is called with an employee whose `ssNum` is already in `d`, its specification says that it must return `duplERR` and leave `d` unchanged. Finally, the requires clause of `uncheckedHire` forbids calling the function with an employee whose `ssNum` is already in `d`.

The only thing of note about `dbase.lh`, Figure 5.15, is that the specification variable and specification type do not appear in it.

An implementation of `dbase` is presented in Appendix B.

A TEST DRIVER FOR DBASE

Before looking at the abstractions used in the implementation of `dbase`, we pause to take a look at some code that uses `dbase`. Figure 5.16 is part of a program we used to test our implementations of the modules specified earlier in this section.

The program `drive` begins with a series of `#includes` of the .h files for the modules containing functions or types that it uses directly. It does not include any subsidiary modules that they may use. While the included .h files are necessary to compile the driver, to understand the code one need look only at the corresponding .lcl files. If the implementation of one of the used modules, such as `empset`, should change, `drive` would have to be re-linked or re-compiled (depending upon whether the .h files `#included` in `drive` were modified), but `drive`'s code would not have to be changed.

After declaring some variables, `drive` initializes the included modules. LCLint will issue a warning if this initialization is not done immediately

```
/*PASS Output from LCL Version 1.7 11-AUG-1992 */
#include "bool.h"
#include "employee.h"
#include "empset.h"
#include "stdio.h"
typedef struct {
    gender g;
    job j;
    int l;
    int h;
} db_q;

typedef enum {
    db_OK,
    salERR,
    genderERR,
    jobERR,
    duplERR,
    missERR} db_status;

extern db_status hire (employee e);

extern void uncheckedHire (employee e);

extern bool fire (int ssNum);

extern int query (db_q q, empset s);

extern bool promote (int ssNum);

extern db_status setSalary (int ssNum, int sal);

extern void db_print (void);

extern void db_initMod (void);
```

FIGURE 5.15. dbase.lh

```c
/* Include those modules that export    */
/* things used explicitly here          */
#include <stdio.h>
#include "bool.h"
#include "employee.h"
#include "empset.h"
#include "dbase.h"

int main(int argc, char *argv[]) {
   employee e;
   empset es;
   char na[10000];
   char * sprintResult;
   int i, j;
   db_status stat;
   db_q q;

/* Initialize the LCL-specified modules */
/* that were included                   */
   bool_initMod();
   employee_initMod();
   empset_initMod();
   db_initMod();

/* Perform tests */
   for (i = 0; i < 20; i++) {
      e.ssNum = i;
      e.salary = 1000 * i;
      if (i < 10) e.gen = MALE; else e.gen = FEMALE;
      if (i < 15) e.j = NONMGR; else e.j = MGR;
      (void) sprintf(na, "J. Doe %d", i);
      employee_setName(&e, na);
      if (i%2 == 0) hire(e);
         else {
            uncheckedHire(e);
            stat = hire(e);
            if (stat != duplERR)
              printf("Error 1: Duplicate not found\n");
         }
   }
   printf("Should print 20 employees:\n");
   db_print();

   /* ... */
```

FIGURE 5.16. Fragment of test driver

following the declarations of the function main. Since the author of main has no way of knowing what modules are used in the implementations of the included modules, the various initMod functions must themselves call the initMod functions of the modules they use. This could result in some initMod functions being called more than once, which is why their specifications typically require them to be idempotent.

The driver then calls some of the specified functions. Effects that are fully constrained by specifications, such as the result returned by fire, are checked internally. Where the specification allows a variety of acceptable effects, output is printed so it can be checked by eye or by a test harness that compares it with the output of a previous run.

We now move down a level of abstraction and specify three interfaces that are useful in implementing the modules specified above. In order to avoid storing more than one copy of an employee, the implementations of db and empset use handles that "point" to objects of type employee. These handles are defined in eref.lcl. The functions specified in ereftab.lcl are used to ensure that the mapping from employees to erefs is one-to-one. The interface erc.lcl exports a type that is basically a bag of erefs. Objects of type erc are used both to represent empsets and within the implementation of db.

EREF

Figure 5.17, introduces an immutable abstract type. Values of type eref can be thought of as abstract pointers to employees. They can be used in much the same way as pointers, except that no functions corresponding to pointer arithmetic have been supplied. Using erefs rather than actual pointers offers several advantages:

- It provides a level of abstraction. The implementor can change the implementation, e.g., from an index into an array to a pointer, without worrying about invalidating client code.

- It allows private storage management. For example, a compacting storage manager can be written, since all access must be through functions in the module.

- It is more general, allowing references to data that is in another address space, on another machine, on a disk, etc.

```
imports employee;

immutable type eref;
spec immutable type map;

spec map m;
constant eref erefNIL = nil;

uses refTable(eref, employee, map);

eref eref_alloc(void) map m; {
   modifies m;
   ensures newInd(result, m^, m');
   }
void eref_free(eref er) map m; {
   requires er ∈ domain(m^);
   modifies m;
   ensures m' = delete(m^, er);
   }
void eref_assign(eref er, employee e) map m; {
   requires er ∈ domain(m^);
   modifies m;
   ensures m' = assign(m^, er, e);
   }
employee eref_get(eref er) map m; {
   requires er ∈ domain(m^);
   ensures result = m^[er];
   }
bool eref_equal(eref er1, eref er2) {
   ensures result = (er1 = er2);
   }
void eref_initMod(void) map m; {
   modifies m;
   ensures m' = new;
   }
```

FIGURE 5.17. eref.lcl

```
refTable(Ind, Val, Tab): trait
    includes Set(Ind, IndSet)
    introduces
        new: → Tab
        assign: Tab, Ind, Val → Tab
        delete: Tab, Ind → Tab
        __[__]: Tab, Ind → Val
        domain: Tab → IndSet
        nil: → Ind
        newInd: Ind, Tab, Tab → Bool
    asserts
      Tab generated by new, assign
      Tab partitioned by __[__], domain
      ∀ i, i1, i2: Ind, v: Val, t,t1,t2: Tab
        delete(new, i) == new;
        delete(assign(t, i1, v), i2) ==
            if i1 = i2
            then delete(t, i2)
            else assign(delete(t, i2), i1, v);
        assign(t, i1, v)[i2] ==
            if i1 = i2 then v else t[i2];
        domain(new) = { };
        domain(assign(t, i, v)) ==
            insert(i, domain(t));
        newInd(i, t1, t2) == ¬ (i ∈ domain(t1))
            ∧ domain(t2) = insert(i, domain(t1))
            ∧ ¬ (i = nil)
```

FIGURE 5.18. refTable.lsl

```
#if !defined(EREF_H)
#define EREF_H

#include "employee.h"

typedef int eref;

/* Private typedefs used in macros  */
typedef enum {used, avail} eref_status;
typedef struct {employee *conts;
                eref_status *status;
                int size;} eref_ERP;

/* Declared here so that macros can use it  */
extern eref_ERP eref_Pool;

#include "eref.lh"

#define erefNIL -1
#define eref_free(er)    (eref_Pool.status[er] = avail)
#define eref_assign(er, e) (eref_Pool.conts[er] = e)
#define eref_get(er)     (eref_Pool.conts[er])
#define eref_equal(er1, er2) (er1 == er2)
#endif
```

FIGURE 5.19. eref.h

```c
#include <stdio.h>
#include "eref.h"

eref_ERP eref_Pool;              /* private */
static bool needsInit = TRUE;    /* private */

eref eref_alloc(void) {
  int i, res;

  for (i=0;
       (eref_Pool.status[i] == used)
          && (i < eref_Pool.size);
       i++);
  res = i;
  if (res == eref_Pool.size) {
    eref_Pool.conts =
      (employee*) realloc(eref_Pool.conts,
                      2*eref_Pool.size*sizeof(employee));

    if (eref_Pool.conts == 0) {
      printf("Malloc returned null in eref_alloc\n");
      exit(1);
    }
    eref_Pool.status =
      (eref_status*) realloc(eref_Pool.status,
                      2*eref_Pool.size*sizeof(eref_status));
    if (eref_Pool.status == 0) {
      printf("Malloc returned null in eref_alloc\n");
      exit(1);
    }
    eref_Pool.size = 2*eref_Pool.size;
    for (i = res+1; i < eref_Pool.size; i++)
        eref_Pool.status[i] = avail;
  }
  eref_Pool.status[res] = used;
  return (eref) res;
}
```

FIGURE 5.20. eref.c, part 1

```
void eref_initMod(void) {
  int i;
  const int size = 16;

  if (needsInit == FALSE) return;
  needsInit = FALSE;
  bool_initMod();
  employee_initMod();
  eref_Pool.conts =
      (employee *) malloc(size*sizeof(employee));
  if (eref_Pool.conts == 0) {
    printf("Malloc returned null in eref_initMod\n");
    exit(1);
  }
  eref_Pool.status =
      (eref_status *) malloc(size*sizeof(eref_status));
  if (eref_Pool.status == 0) {
    printf("Malloc returned null in eref_initMod\n");
    exit(1);
  }
  eref_Pool.size = size;
  for (i = 0; i < size; i++) eref_Pool.status[i] = avail;
}
```

FIGURE 5.20. eref.c, part 2

In eref.lcl, the specification variable m is used to keep track of the set of extant erefs (so that the specification eref_alloc can say that the result is a new eref) and of the mapping from erefs to employees. Trait refTable, Figure 5.18, specifies the operators on the values of objects of type map.

Figures 5.19 and 5.20 contain an implementation of eref.

The implementation variable eref_Pool has much the same role in the implementation as the specification variable m did in eref.lcl. However, there are other implementations that would not have anything corresponding to m, for example, one that used the C type employee * to represent erefs. Because the implementation variable eref_Pool is used in macro definitions, C requires it to be declared extern in eref.h, even though clients of eref should not reference it—or even know about its existence.

In this implementation of eref, the function eref_equal is implemented by a macro that uses ==. However, one can imagine implementations of eref for which this would not work. Suppose, for example, the implementation used a gratuitous level of indirection and made int * the representation of eref. Then eref_equal would have to be implemented as *er1 == *er2. This illustrates why LCLint will generate a warning if clients use the == operator directly, rather than calling eref_equal.

ERC

Figures 5.21 and 5.22 together specify a set of functions operating on the mutable abstract types, erc (for "employee ref collection") and ercIter. These types and functions are used in the implementation of both empset and dbase.

An erc is essentially a bag.[9] Most of the functions on erc's are unremarkable.; the unusual functions in this specification are those that deal with ercIters.

Objects of type ercIter are used by clients to iterate over all the elements of an erc. In the specification, Figure 5.21, (though not in the implementation, discussed on page 99) ercIter's are modeled as a pair consisting of the erc to be iterated over and a bag containing those erefs that have not been yielded to the client. The function erc_iterStart

---

[9]Trait Bag can be found in Appendix A, page 169.

```
erc: trait
  assumes CTrait
  includes Bag(eref, ercElems)
  erc tuple of val:ercElems, activeIters: int
  ercIter tuple of toYield: ercElems, eObj: ercObj
  introduces
    { }: → erc
    yielded: eref, ercIter, ercIter → bool
    startIter: erc → erc
    endIter: erc → erc
  asserts
    ∀ e: eref, it1, it2: ercIter, c: erc
      { } == [{ }, 0];
      yielded(e, it1, it2) == e ∈ it1.toYield
           ∧ it2 = [delete(e, it1.toYield), it1.eObj];
      startIter(c) == [c.val, c.activeIters + 1];
      endIter(c) == [c.val, c.activeIters - 1]
```

FIGURE 5.21. erc.lsl

```
imports eref;

mutable type erc;
mutable type ercIter;

uses erc(obj erc for ercObj), sprint(erc, char[]);

erc erc_create(void) {
    ensures fresh(result) ∧ result' = { };
    }
void erc_clear(erc c) {
    requires c^.activeIters = 0;
    modifies c;
    ensures c' = { };
    }
void erc_insert(erc c, eref er) {
    requires c^.activeIters = 0 ∧ er ≠ erefNIL;
    modifies c;
    ensures c' = [insert(er, c^.val), 0];
    }
bool erc_delete(erc c, eref er) {
    requires c^.activeIters = 0;
    modifies c;
    ensures result = er ∈ c^.val
        ∧ c' = [delete(er, c^.val), 0];
    }
bool erc_member(eref er, erc c) {
    ensures result = er ∈ c^.val;
    }
```

FIGURE 5.22. erc.lcl, part 1

```
eref erc_choose(erc c) {
   requires size(c^.val) ≠ 0;
   ensures result ∈ c^.val;
   }
int erc_size(erc c) {
   ensures result = size(c^.val);
   }
ercIter erc_iterStart(erc c) {
   modifies c;
   ensures fresh(result) ∧ result' = [c^.val, c]
     ∧ c' = startIter(c^);
   }
eref erc_yield(ercIter it) {
   modifies it, it^.eObj;
   ensures if it^.toYield ≠ { }
     then yielded(result, it^, it')
       ∧ (it^.eObj)' = (it^.eObj)^
     else result = erefNIL ∧ trashed(it)
       ∧ (it^.eObj)' = endIter((it^.eObj)^);
   }
void erc_iterFinal(ercIter it) {
   modifies it, it^.eObj;
   ensures trashed(it)
     ∧ (it^.eObj)' = endIter((it^.eObj)^);
   }
void erc_join(erc c1, erc c2) {
   requires c1^.activeIters = 0;
   modifies c1;
   ensures c1' = [c1^.val ∪ c2^.val, 0];
   }
char *erc_sprint(erc c) {
   ensures isSprint(result[]', c^) ∧ fresh(result[]);
   }
void erc_final(erc c) {
   modifies c;
   ensures trashed(c);
   }
void erc_initMod(void) {
   ensures true;
   }
```

FIGURE 5.22. erc.lcl, part 2

maps an `erc` into an `ercIter` in which all the elements remain to be yielded. Each time `erc_yield` is called with this object, it returns an `eref` and updates the `ercIter` by deleting the returned `eref` from the bag of `eref`s that remain to be yielded. When each `eref` has been yielded as many times as it occurs in the `erc`, `erc_yield` returns `erefNIL`.

Iterator functions are typically used in code of the form

```
eref er;
erc c;
ercIter it;
. . .
for_ercElems(er, it, c) {
  Body of loop
  }
```

where `for_ercElems` is defined by the macro

```
#define for_ercElems(er, it, c)\
  for (er = erc_yield(it = erc_iterStart(c));\
       er != erefNIL;\
       er = erc_yield(it))
```

It is often the case that the body of an iteration itself uses an iterator. The introduction of `ercIter`s makes it possible to have nested iterations over the same `erc`.

One question that arises with this programming paradigm concerns what happens if the `erc` is modified within the body of the loop. Writing specifications that give a precise semantics for such a situation is not difficult. However, building an efficient implementation is. For that reason, our specification forbids modification of an `erc` that is being iterated over.

In `erc.lsl`, Figure 5.21, an `erc` is modeled as a pair of a bag of `ercElems` and an `int`. The bag is used to contain the elements of the `erc` and the `int` is used, in `erc.lcl`, to keep track of the number of active iterators. This makes it possible to write requires clauses that prohibit calling a function that might modify an `erc` while that `erc` is being iterated over. Conceptually, the function `erc_iterStart` increments the number of active iterators. The function `erc_yield` decrements the number of active iterators when it has yielded the last element. The function `erc_iterFinal` also decrements the number of active iterators. This function should be called before exiting prematurely (e.g., by `break` or `return`) from the body of an iteration.[10]

---

[10]The function `erc_iterFinal` can be used in macros to define versions of `return` and `break` that are appropriate for use within iterations. An example of this appears in Appendix B.

```
imports employee, eref;

mutable type ereftab;

uses ereftab;

ereftab ereftab_create(void) {
   ensures result' = empty;
   }
void ereftab_insert(ereftab t, employee e, eref er) {
   requires getERef(t^, e) = erefNIL;
   modifies t;
   ensures t' = add(t^, e, er);
   }
bool ereftab_delete(ereftab t, eref er) {
   modifies t;
   ensures result = in(t^, er) ∧ t' = delete(t^, er);
   }
eref ereftab_lookup(employee e, ereftab t) {
   ensures result = getERef(t^, e);
   }
void ereftab_initMod(void) {
    ensures true;
    }
```

FIGURE 5.23. ereftab.lcl

Again, the implementation is not presented here, but appears in Appendix B.

EREFTAB

The last module in our example is ereftab, Figures 5.23 and 5.24. It is used to create a one-to-one mapping from employees to erefs. It makes it unnecessary to store multiple copies of the same employee record within the implementation of empset.

The intended use of ereftab_insert is to put an employee in an ereftab only after a lookup has failed to find an eref for that employee. The requires clause of ereftab_insert formalizes this property, and allows the implementation not to duplicate a test that has just been made by the client.

The implementation of ereftab is unremarkable, and is not presented.

```
ereftab: trait
  assumes CTrait
  introduces
    empty: → ereftab
    add: ereftab, employee, eref → ereftab
    delete: ereftab, eref → ereftab
    getERef: ereftab, employee → eref
    erefNIL: → eref
    in: ereftab, eref → bool
    size: ereftab → int
  asserts
    ereftab generated by empty, add
    ereftab partitioned by getERef
    ∀ e, e1: employee, er, er1: eref, t: ereftab
      delete(empty, er) == empty;
      delete(add(t, e, er), er1) ==
          if er = er1 then t
          else add(delete(t, er1), e, er);
      in(empty, er) == false;
      in(add(t, e, er), er1) == er = er1 ∨ in(t, er);
      getERef(empty, e1) == erefNIL;
      getERef(add(t, e, er), e1) ==
          if e = e1 then er else getERef(t, e1);
      size(empty) == 0;
      size(add(t, e, er)) ==
          1 + (if in(t, er) then 0 else 1)
```

FIGURE 5.24. ereftab.lsl

```
typedef struct _elem
        {eref val; struct _elem *next;} ercElem;
typedef ercElem *ercList;
typedef struct {ercList vals; int size;} ercInfo;
typedef ercInfo *erc;
typedef ercList *ercIter;
```

FIGURE 5.25. erc's representation

IMPLEMENTATION NOTES

Here we take the opportunity to make some comments about the relationship of these specifications to the implementations presented in Appendix B.

In erc.lcl, Figure 5.22, the value of an object of type erc was modeled as a pair of a bag and an integer. The integer was used to keep track of the number of active iterators. Figure 5.25 contains the representation used in the implementation of erc and ercIter. The representation of erc is a pair, but the integer is not used to keep track of the number of active iterators. Rather it contains the number of elements in the erc. In fact, the implementation has no need to keep track of the number of active iterators. It is the responsibility of the clients of this interface to ensure that the requires clause holds whenever a function is called. It might be an appropriate application of defensive programming for the implementor of erc.lcl to keep track of the number of active iterators and check that requires clauses hold on entry to functions, but it is not required by the specification.

The implementation of empset uses an erc to represent an empset. (Recall that the val field of an erc is a bag.) The implementation also uses a non-exported module-level variable, known, to avoid allocating space for the same employee more than once. The first time an employee is inserted into any empset, it is also inserted into known and a newly allocated eref is inserted into the erc. On subsequent inserts of the same employee into any empset, the old eref is reused. This auxiliary data structure is shared by the implementation of all objects of type empset, but this sharing is not visible to clients.

Figure 5.26 contains a representation invariant for the implementation of empset. The implementation ensures that this invariant is established by empset_create and preserved by all other functions in the empset interface. The first conjunct of the invariant asserts that no eref occurs

```
∀ s:empset
  (∀ er:eref (count(er, s.val) ≤ 1)
    ∧ s.activeIters = 0
    ∧ ∀ er:eref
      (count(er, s.val) = 1 ⇒ er ∈ known)
```

FIGURE 5.26. Representation invariant for empset

```
#define firstERC mMGRS
#define lastERC fNON
#define numERCS (lastERC - firstERC + 1)

typedef enum {mMGRS, fMGRS, mNON, fNON} employeeKinds;

erc db[numERCS];
```

FIGURE 5.27. dbase.c fragment

more than once in the val part of an erc used to represent an empset. The second conjunct corresponds to the requires clause of many of the functions of type erc, and therefore must be maintained so that the implementation of empset can use those functions. The third conjunct gives a relationship that must always hold between the module specification variable known and any erc representing an empset.

The implementation of dbase is considerably longer than that of the other modules specified here. It is also somewhat different in structure. Unlike empset.h and erc.h, dbase.h contains no typedef (although it does inherit typedefs of exposed types from dbase.lh). This is because dbase.lcl exports no abstract types and the implementation of dbase doesn't use any macros that depend on locally defined types. Information pertinent only to compiling the implementation itself is restricted to dbase.c, Figure 5.27.

The specification variable d in dbase.lcl is implemented by the variable db. We chose a different name for the variable in the implementation to emphasize that there is no necessary correspondence between module-level variables appearing in the implementation and specification variables appearing in the specification. It is purely accidental that our specification variable corresponds to a single implementation variable; one of our earlier implementations of the interface used four distinct ercs to represent d.

The correctness of the implementations of the functions in dbase.c depends upon the maintenance of the representation invariant given in Figure 5.26. That this holds can be shown by an inductive argument:

- It is established by dbase_initMod.

- For each exported function, if the invariant and the requires clause hold on entry, the invariant will hold upon termination. In discharging this step of the proof, it is necessary to examine even those functions whose specification does not allow them to modify d, since they might still modify the representation of d, i.e., the array db.

The implementation of dbase includes several functions that do not appear in dbase.lcl and therefore are not accessible to clients. It would be acceptable for these functions to break the invariant temporarily (although, in fact, they don't).

Chapter 6

LM3: A Larch Interface Language for Modula-3

This chapter describes much of LM3, version 1.1, and gives an informal description of its semantics. It skims somewhat rapidly over the role of the LSL specification tier, which is quite similar to that for LCL, as discussed in the previous chapter.

Because Modula-3 is structured around the definition and use of explicit interfaces, LM3 specifications are more intimately related to Modula-3 programs than LCL specifications are to C programs. Because Modula-3's opaque types and revelations provide direct support for abstract types, LM3 doesn't need to add much in that area. Because Modula-3's REF types are more disciplined than C's pointer types, LM3's storage model is somewhat simpler. Because Modula-3 provides garbage collection, specifications don't have to say as much about storage management; for example, there is no need for anything corresponding to trashed in LCL. But subtyping and concurrency raise issues that make LM3 complicated in other ways.

LM3 provides constructs for specifying:

- types, both fully exposed and abstract (opaque);

- procedures and object methods (collectively, *routines*);

- invariants, for both types and modules;

- concurrency and synchronization.

This chapter is intended for Modula-3 programmers—practicing or potential. We assume some familiarity with Modula-3. If you are not acquainted with Modula-3, you may wish to consult a Modula-3 text [52, 69].

6.1   The relation between LM3 and Modula-3

Modula-3 has well-defined notions of interface and implementation:

- An *interface file* (.i3 file or .ig file) declares the components of the module's interface and documents the intended uses of exported types and the actions of exported procedures. LM3 specifications are incorporated in interface files; we will often call such augmented files *interface specifications.*

- An *implementation file* (.m3 file or .mg file) supplies the representations of the types and the bodies of the procedures and object methods declared in the interface, as well as code that is private to the module.

Clients of a module should look at its interface, not its implementation. LM3 is used to provide clients with a precise description of the functionality of the interface.[1] An LM3 specification also provides implementors a contract with precise information about what they are to implement.

There are two kinds of information in an LM3 interface specification:

- Modula-3 declarations. Each Modula-3 interface file is also an LM3 specification. There is a built-in association of Modula-3 base types and type constructors with LSL sorts, and there is a standard set of traits for Modula-3 that provides operators on these sorts.

- LM3 pragmas. As will be discussed in the rest of this chapter, LM3 annotations are incorporated in Modula-3 as *pragmas*, set off by the brackets <* and *>. Pragmas embedded in interface files can introduce abstract types and give constraints on types, variables, and routines. Since the compiler ignores pragmas that it does not recognize, they provide a convenient way of embedding specification information in the program text. LM3 annotations may be thought of as formalized comments within the interface file.

6.2   The LM3 semantic model

The LM3 and LSL tiers of a specification are connected as described in Chapter 3. LM3 annotations are written using LSL terms plus some syntactic sugar to make specifications more Modula-3-like in appearance.

Since LM3, like Modula-3, is statically typed, the kind of values that a variable can contain in any state is fixed. Similarly, each LSL value has a unique *sort.* To connect the two languages, there is a mapping from LM3

---

[1]LM3 is also used to annotate implementations for program verification. This aspect of LM3 is not addressed in this book.

types to LSL sorts. Each built-in type of Modula-3, each type built from Modula-3 type constructors (e.g., ARRAY [1..100] OF INTEGER), and each abstract type defined in LM3 is *based on* an LSL sort. The sort on which a type is based does not appear explicitly in LM3 specifications. Instead, an LM3 type name or other type expression stands for its associated sort. LM3 follows Modula-3's type checking rules [69].

Standard LSL traits define operators of the sorts upon which Modula-3 built-in types (e.g., INTEGER and TEXT) are based. Users familiar with Modula-3 will already have some intuition about these operators. Specifier-supplied traits are used to introduce application-specific operators. A *traits clause* is used to incorporate specifier-supplied traits and to connect user-defined types to LSL sorts.

An LM3 interface specification defines the functional behavior of a collection of exported *routines* (procedures and methods), variables, and constants. From a semantic point of view, there is no significant difference between procedures and methods; methods are just procedures with an implicit SELF parameter and a slightly different syntax.

A routine may communicate with its callers by returning a result, by accessing variables accessible to the caller, by modifying such variables, or by raising an exception. The specification of each routine in an interface can be studied, understood, and used without reference to the specifications of other routines.

Each routine is specified by a predicate on a pair of *states*—the *pre-state* and the *post-state*—that defines the set of state transformations (actions) the routine is allowed to perform.[2]

A state is a repository for entities that can be changed by routines. It is a mapping from *entire variables* to *values*. Each program and specification variable is a coordinate of the state space; entire variables are the orthogonal coordinates. Each entire variable can be assigned a value without affecting the value of any other entire variable. For example, if A is an array variable, A is entire, but A[i] and A[j] are not, since assigning to one might change the other, depending on the values of i and j. Each field of an object type is an entire variable, indexed by objects. However, t.f is analogous to f[t] and is not entire.

- The *global state* for an interface specification is defined by its type, variable and constant declarations, and the global states of the

---

[2]In our discussion of concurrency, we will generalize the predicate to apply to a sequence of pairs of states, rather than just a single pair; see page 116.

interfaces it imports. It may include auxiliary variables and fields introduced in pragmas purely for the purposes of the specification.

- The *local state* for a routine specification is given by its formal parameter list, RESULT (which represents the returned value, if any), RAISEVAL (which represents the normal or exceptional outcome), RAISEARG (which represents the value of the argument to RAISE, if any), CURRENT (which represents the identity of the thread that called the routine), and the components of the global state that the routine is allowed to access.

- The *target variables* of a routine are those variables to which it is allowed to assign new values. They are a subset of its local state, and are explicitly listed in its specification.

- Within a specification, an *immutable value* (constant) is represented directly by its name. The value of a *variable* in the pre-state is also represented by its name; the value of a target variable in the post-state is represented by its name followed by a prime (′).

As discussed in Chapter 3, a routine specification consists of a *routine declaration* augmented by a body containing REQUIRES, MODIFIES, and ENSURES clauses. It effectively separates the obligations of clients and implementations. The requires clause gives the obligations of the client, which the implementor is entitled to presume. The modifies and ensures clauses give the obligations of the implementor, which (along with termination) the client is entitled to presume.

6.3   A guided tour through an LM3 specification

To show how LM3 is used, we present and discuss an example that makes use of most of its features. The example is only superficially realistic; it was structured to use language constructs in the order that we want to discuss them.

AN EMPLOYEE DATABASE

Our example is a simple database that holds information about employees. If you have already looked at the example in Chapter 5, you should note that this is not the same design. For example, this database stores sharable Employee objects; Chapter 5's database stores values of records about employees. Some of the differences are due to differences between the styles that are natural in C and in Modula-3; some are arbitrary.

An interesting feature of this database is that these routines may be invoked concurrently and therefore require mutually exclusive access to the shared data. How this mutual exclusion is ensured is up to the implementation; the specification does not say. However, it does say which routines are allowed to be non-atomic; all the rest must appear atomic to their users.

We start with simple interfaces and build up to more complex ones:

- `EmployeeData` contains no specification pragmas, but shows how Modula-3 declarations are interpreted as LM3 specifications.

- `Employee` introduces some explicit LM3 type specifications and illustrates the specification of methods of an exposed type.

- `GenericSet` specifies a type that is generic and opaque, and has a nondeterministic method.

- `EmployeeSet` shows the instantiation of a generic interface.

- `EmployeeGroup` illustrates simple subtyping and a non-atomic routine.

- `EmployeeDatabase` uses a combination of previously-discussed features.

- `EmployeeSetFriends` illustrates the use of a partial revelation to give access to part of the representation of an abstract type.

We specify each interface, and describe the meaning of the LM3 constructs it introduces.

EMPLOYEEDATA

The interface specification in Figure 6.1 declares some simple types that we use in later interfaces, but contains no specification pragmas. It illustrates *exposed types*, whose full specification is given by the semantics of their Modula-3 declarations. The specification states that:

- `MaxSal` is a constant of sort `Int`.

```
INTERFACE EmployeeData;
  CONST MaxSal = 1000000;
  TYPE
    Gender = {Male, Female};
    Job = {NonMgr, Manager};
    Salary = [1 .. MaxSal];
    SSnum = INTEGER;
END EmployeeData.
```

FIGURE 6.1. EmployeeData.i3

- In any state, the value of a variable of type Salary has sort Int from the trait Integer, (since INTEGER is the base type of any integer subrange). Furthermore, the value will be between 1 and MaxSal. SSnum is simply a renaming of INTEGER.

- Gender and Job are enumeration types with the constants Gender.Male and Gender.Female and Job.NonMgr and Job.Manager. These constants may be used in specifications just as they are in programs.

EMPLOYEE

Figure 6.2 defines a data type used to hold information about individual employees.

The *imports clause* of Employee says that its interface specification depends on EmployeeData's interface specification; it gives Employee and its clients access to the constants, variables, types, and routines specified in EmployeeData. It also makes the trait associated with EmployeeData a part of Employee's associated trait. This specification dependency should not be confused with an implementation dependency, where an interface is used within the implementation of a module.

Following a common convention in Modula-3, the principal type of the Employee interface is named T, for easy reference within the interface specification and implementation. Outside the module, it is referred to as Employee.T.

T is an exposed *object type*.[3] It doesn't introduce any abstraction, and

---

[3] Unfortunately, "object" means different things in different programming languages. In Modula-3, an object type is an explicit reference type with *fields* and *methods*.

```
INTERFACE Employee;
  IMPORT EmployeeData;
  TYPE
    T = OBJECT
        ssnum : EmployeeData.SSnum;
        name  : TEXT;
        salary: EmployeeData.Salary;
        gender: EmployeeData.Gender;
        job   : EmployeeData.Job;
      METHODS
        promote (increase: EmployeeData.Salary)
            RAISES {AlreadyManager};
    END;
  EXCEPTION AlreadyManager;
<* METHOD T.promote(increase)
    REQUIRES
      (SELF.salary + increase) ≤ EmployeeData.MaxSal
    MODIFIES SELF.job, SELF.salary
    ENSURES SELF.job′ = EmployeeData.Job.Manager
      ∧ SELF.salary′ = SELF.salary + increase
    EXCEPT SELF.job = EmployeeData.Job.Manager
      => (RAISEVAL = AlreadyManager ∧ UNCHANGED(ALL))
*>
END Employee.
```

FIGURE 6.2. Employee.i3

its data representation is fully defined by Modula-3. The implicit operators for an object type allow access to its fields, so that, for example, t.name' refers to the value in the post-state of the name field of the object t.

This interface provides our first example of a specification that goes beyond what is provided by Modula-3 itself. In the specification of the method T.promote

- the requires clause says that promote should be called with a value of increase that results in a valid raise; the raise will be positive because of the type of increase. If the raise is too big, the behavior of promote is unconstrained.

- the modifies clause (target list) says that promote may not alter the values of any client-visible variables except the object's own job and salary fields.

- the ensures clause says that promote must change the job and salary fields in particular ways. This postcondition is written in two parts:

  - The first part describes the *normal* result of an invocation of promote: the job field will be changed to Manager and the salary field will be incremented by increase.

  - The second part describes the *exceptional* behavior. If SELF.job is already Manager then promote must raise the exception AlreadyManager and change nothing.

There are several more things to note about the constructs used in this specification:

- An *except clause* consists of one or more guarded predicates. If any *guard* (a predicate before =>) is true, then the method must ensure the postcondition given after one of the true guards, rather than the normal postcondition. If more than one guard is true, the implementation may satisfy any of the associated postconditions, nondeterministically.

- RAISEVAL is a special component of the state; a value other than RETURNS in the post-state represents the raising of an exception. If there is no except clause, RAISEVAL = RETURNS is implicit.

- We do not follow RESULT and RAISEVAL with primes; since they are meaningful only in the post-state, there is no ambiguity.

- The UNCHANGED operator is a shorthand for saying that the values of a list of variables may not change between the pre-state and the post-state, even though they are in the target list. It is equivalent to saying $x' = x$ for each x in the list. ALL is a further shorthand for the complete target list.

GENERICSET

The interface in Figure 6.3 provides a generic set abstraction. This is our first type that is not exposed. T is an *abstract type* whose representation is hidden from clients. In Modula-3, this is called an *opaque type*.

T <: Public says that T is a *subtype* of the type Public. It is a common convention in Modula-3 to use an auxiliary type named Public to declare the methods and fields exported by an opaque type.

Since we have chosen not to make the representation of T visible, we have to provide some way to represent its values in specifications. We declare a *specification field*, T.set to denote the value represented by the hidden components. Within the specification, we treat it as though it

```
GENERIC INTERFACE GenericSet(E);
  EXCEPTION NotFound;
  TYPE
    T <: Public;
    Public = OBJECT
      METHODS
        init           ();
        copyTo         (newCopy: T);
        freshCopy      (): T;
        size           (): CARDINAL;
        insert         (e: E.T);
        remove         (e: E.T);
        union          (s: T);
        disjointUnion  (s: T);
        intersect      (s: T);
        member         (e: E.T): BOOLEAN;
        choose         (): E.T;
  END;
```

FIGURE 6.3. GenericSet.ig, part 1

```
< * TRAITS Set(E.T FOR E, ETSet FOR C);
    TYPE ETSet;
    FIELDS OF T set: ETSet
  METHOD T.init()
    MODIFIES SELF.set
    ENSURES SELF.set' = { }
  METHOD T.copyTo(newCopy)
    MODIFIES newCopy.set
    ENSURES newCopy.set' = SELF.set
  METHOD T.freshCopy()
    ENSURES RESULT.set' = SELF.set ∧ FRESH(RESULT)
  METHOD T.size()
    ENSURES RESULT = size(SELF.set)
  METHOD T.insert(e)
    MODIFIES SELF.set
    ENSURES SELF.set' = insert(e, SELF.set)
  METHOD T.remove(e)
    MODIFIES SELF.set
    ENSURES SELF.set' = delete(e, SELF.set)
  METHOD T.union(s)
    MODIFIES SELF.set
    ENSURES SELF.set' = SELF.set ∪ s.set
  METHOD T.disjointUnion(s)
    REQUIRES SELF.set ∩ s.set = { }
    MODIFIES SELF.set
    ENSURES SELF.set' = SELF.set ∪ s.set
  METHOD T.intersect(s)
    MODIFIES SELF.set
    ENSURES SELF.set' = SELF.set ∩ s.set
  METHOD T.member(e)
    ENSURES RESULT = e ∈ SELF.set
  METHOD T.choose()
    REQUIRES SELF.set ≠ { }
    MODIFIES SELF.set
    ENSURES RESULT ∈ SELF.set
        ∧ SELF.set' = delete(RESULT, SELF.set)
*>
END GenericSet.
```

FIGURE 6.3. GenericSet.ig, part 2

```
INTERFACE EmployeeSet = GenericSet(Employee)
END EmployeeSet.
```

FIGURE 6.4. EmployeeSet.i3

were declared as an ordinary field of T. We don't have to include it in the implementation, but any *revelation* of hidden fields of T must have an associated *abstraction relation* that shows how the specification and implementation values are related.

In our earlier examples, the trait associated with each interface has been implicit, entirely composed of built-in traits associated with Modula-3 and with the types and type constructors appearing in declarations. Here, the *traits clause* explicitly includes Set, page 167, into the trait associated with the GenericSet interface, renaming the formal parameters of the trait to the sorts on which the types E.T and ETSet are based. E.T is a program type, and ETSet is a *specification type*, introduced in this pragma as the type for the specification field, set.

Most of the method specifications follow the same pattern as our previous example, using the specification fields of T rather than actual fields. T.init, for example, ensures that the abstract field SELF.set has the value {} when it returns. The specification of T.copyTo ensures that the set field of the object passed in as a parameter becomes equal to SELF.set. This is quite different from saying that SELF is assigned to a VAR parameter, which would be specified as follows:

```
METHOD T.assign(target)
  MODIFIES target
  ENSURES target' = SELF
```

The choose method is an example of a specification of a non-deterministic routine. The method is required to remove and return some value from the set. No information is given about which element is to be chosen; the implementation may use this freedom to improve efficiency, so clients must not rely on any particular choice.

EMPLOYEESET

EmployeeSet is a simple interface that instantiates the interface GenericSet passing the Employee interface for the formal parameter E. The instantiated program type EmployeeSet.T has an instantiated specification field set with instantiated specification type EmployeeSet.ETSet that holds a set of Employee.Ts.

EMPLOYEEGROUP

Figure 6.5 introduces a specialization of EmployeeSet that has an extra component, manager. Informally, a group is a set of employees with one distinguished member. The only extra operation we add to a group is a method to make an employee (who may or may not already be a member of the group) the manager of the group.

In this interface, we illustrate the interaction between specification and subtyping, show a type invariant, and specify a non-atomic method.

Here, we have a *partially opaque* type. The type has one visible field, EmployeeGroup.T.manager, but there may also be hidden fields used by the implementation. Since T is a subtype of EmployeeSet.T, both the exposed and specification fields of EmployeeSet.T can be used in the specification of T. We use the local manager field and the inherited set specification field.

```
INTERFACE EmployeeGroup;
  IMPORT EmployeeData, Employee, EmployeeSet;
  TYPE
    T <: Public;
    Public = EmployeeSet.T OBJECT
        manager: Employee.T;
      METHODS
        copyTo (newCopy: T);
        freshCopy(): T;
        makeManager(e: Employee.T);
      END;
  PROCEDURE Subordinates (t: T): EmployeeSet.T;
<* TYPE_INVARIANT t: T
    t.manager = NIL
      V (t.manager.job = EmployeeData.Job.Manager
          ∧ t.manager ∈ t.set)
  PROCEDURE Subordinates(t)
    ENSURES RESULT.set' = delete(t.manager, t.set)
      ∧ FRESH(RESULT)
*>
```

FIGURE 6.5. EmployeeGroup.i3, part 1

```
<* STRENGTHEN T.init()
    MODIFIES SELF.manager
    ENSURES SELF.manager = NIL
  STRENGTHEN T.remove(e)
    MODIFIES SELF.manager
    ENSURES IF e = SELF.manager
            THEN SELF.manager' = NIL
            ELSE UNCHANGED(SELF.manager)
  STRENGTHEN T.intersect(s)
    MODIFIES SELF.manager
    ENSURES IF SELF.manager ∈ SELF.set'
            THEN UNCHANGED(SELF.manager)
            ELSE SELF.manager = NIL
  STRENGTHEN T.choose()
    MODIFIES SELF.manager
    ENSURES IF RESULT = SELF.manager
            THEN SELF.manager = NIL
            ELSE UNCHANGED(SELF.manager)
  METHOD T.copyTo(newCopy)
    MODIFIES newCopy.manager, newCopy.set
    ENSURES newCopy.manager' = SELF.manager
      ∧ newCopy.set' = SELF.set
  METHOD T.freshCopy()
    ENSURES RESULT.manager = SELF.manager
      ∧   RESULT.set = SELF.set
      ∧   FRESH(RESULT)
  METHOD T.makeManager(e)
    MODIFIES e.job, SELF.manager, SELF.set
    COMPOSITION OF promote; add_to_group; install END
      ACTION promote
        ENSURES e.job' = EmployeeData.Job.Manager
          ∧ UNCHANGED(SELF.manager, SELF.set)
      ACTION add_to_group
        ENSURES SELF.set' = insert(e, SELF.set)
          ∧ UNCHANGED(e.job, SELF.manager)
      ACTION install
        ENSURES SELF.manager' = e
          ∧ UNCHANGED(e.job, SELF.set)
*>
END EmployeeGroup.
```

FIGURE 6.5. EmployeeGroup.i3, part 2

The first new construct in this specification is a *type invariant*. The meaning of this clause is that, in any state visible to a client, each instance of T either has no `manager` or has a `manager` field whose `job` field has the value `Manager` and that `manager` will always be a member of its `set`. This invariant is conjoined to the precondition of each routine and action in the interface that may read something of type T, and to the postcondition of each routine and action in the interface that may modify or return something of type T. Variable names in the invariant are implicitly primed for postconditions.

The procedure in the interface, `subordinates`, returns the members of a given group, excluding the manager. It could, of course, have been specified as a method on T, but it is also perfectly valid to do it this way. The only item of interest in the specification of is the use of the `set` field of a T as a value for the equivalent field in an `EmployeeSet.T`. This is permitted since the `set` field was inherited from the supertype.

Each of the methods that T inherits from `EmployeeSet.T` has an inherited specification. A subtype method always inherits the specification of the corresponding method for the supertype; otherwise it would not be sensible to use values of the subtype in contexts where values of the supertype is expected. Since the subtype is more specialized, it is often appropriate to give it a stronger specification. This is done using a *strengthen clause*.

For example, because an `EmployeeGroup.T` has a `manager` field, and an `EmployeeSet.T` does not, most methods that modify values of type `EmployeeGroup.T` should have strengthened specifications.

For some of the methods, such as `size`, `manager` is simply irrelevant. Generalizing the principle that an omitted modifies clause means `modifies nothing`, the absence of any further specification of `size` means that it leaves the `manager` field unchanged. This interpretation also suffices for the specifications of `insert`, `union`, `disjointUnion`, and `member`.

The incremental specifications of `init`, `remove`, `intersect` and `choose` are simple: they just say what value `manager` is to have in the post-state. This extra clause is conjoined onto the specification inherited from the supertype.

The treatment of the methods `copyTo` and `freshCopy` is more complex, but not unusual. In Modula-3, only the implicit `SELF` parameter to a method gets the subtype by inheritance. So both inherited methods produce an `EmployeeSet.T`, rather than an `EmployeeGroup.T`. To get around this, we use a standard trick. We introduce new methods that produce EmployeeGroup.T values, intentionally giving them the same names, so they obscure the inherited methods. Since the new methods have different signatures, STRENGTHEN is not appropriate and we give full specifications for the new methods.[4]

The method makeManager introduces another feature. Modula-3 has built-in support for *threads*, which are lightweight units of concurrency that may share a state space. When specifying routines that may be called from multiple threads, we have to be concerned about the possibility of interference among these threads. LM3 provides constructs to specify each non-atomic routine as a sequence of *atomic actions*.[5] To clients of an interface, atomic actions must always appear to have executed in some particular order; any concurrency in the implementation must be hidden.

The behavior of makeManager is specified as three atomic actions. Consider the elements of its specification:

- The modifies clause is the same as for an atomic routine. It restricts each of the actions to a subset of its target list. None of the actions can modify non-target variables. An action specification may further limit the changes to a subset of the target list, by indicating components that are not to be changed by that action.

- Rather than a single ensures clause, the method is specified as a *composition* of a three *actions*. Each action has an associated specification that can be read as if it were a routine specification without a requires clause:

  1. promote—must change the job component of the e parameter to Manager, and must not alter either the manager field or the set specification field;

  2. add_to_group—must insert e into the group;

  3. install—must make e the manager of the group.

---

[4] The signature of inherited methods sometimes confuses novice Modula-3 programmers and they make the mistake of expecting all parameters of the supertype to be converted to the subtype. The same misunderstanding will lead to the detectable mistake of using STRENGTHEN when it is inappropriate.

[5] This section only touches the tip of the concurrency iceberg. It does not discuss synchronization operations or the general case where routines may have action sequences of arbitrary length. A more complete example is contained in Chapter 5 of *Systems Programming with Modula-3* [69], which uses an earlier version of LM3 to specify Modula-3's synchronization primitives.

So the overall effect of the method is to make e the manager of the group, while ensuring that each action preserves the invariant on T. Preserving the invariant between actions is important because other actions might be interleaved between promote and add_to_group or between add_to_group and install.

EMPLOYEEDATABASE

An EmployeeDatabase, Figure 6.6, provides a collection of routines, including both queries and updates, over a set of EmployeeGroups and their employees.

```
INTERFACE EmployeeDatabase;
  IMPORT EmployeeData, Employee, EmployeeSet,
    EmployeeGroup;
  TYPE
    T <: Public;
    Public = OBJECT
      METHODS
        init ();
        query (q: Query): EmployeeSet.T;
        hire (e: Employee.T; g: EmployeeGroup.T)
          RAISES {AlreadyEmployee};
        getGroup (e: Employee.T): EmployeeGroup.T
          RAISES {NotEmployee};
        createGroup (man: Employee.T):
          EmployeeGroup.T;
        removeGroup (g: EmployeeGroup.T);
      END;
    Query = RECORD
      g := EmployeeData.Gender.Male;
      j := EmployeeData.Job.NonMgr;
      testGender, testJob: BOOLEAN := FALSE;
      low := FIRST(EmployeeData.Salary);
      high := LAST(EmployeeData.Salary);
    END;
  EXCEPTION AlreadyEmployee;
  EXCEPTION NotEmployee;
```

FIGURE 6.6. EmployeeDatabase.i3, part 1

```
< * TRAITS Set (EmployeeGroup.T FOR E, EGSet FOR C);
   TYPE EGSet;
   FIELDS OF T  set: EGSet;
   METHOD T.init
     MODIFIES SELF.set
     ENSURES SELF.set' = { }
   METHOD T.query(q)
     ENSURES ∀ e:Employee.T;
         e ∈ RESULT.set'
            ⇔
         (∃ gr:EmployeeGroup.T;
           gr ∈ SELF.set
             ∧ e ∈ gr.set
             ∧ (q.testGender ⇒ q.g = e.gender)
             ∧ (q.testJob ⇒ q.j = e.job)
             ∧ q.low ≤ e.salary
             ∧ e.salary ≤ q.high)
   METHOD T.hire(e, g)
     REQUIRES g ∈ SELF.set
     MODIFIES g.set
     ENSURES g.set' = insert(e, g.set)
     EXCEPT ∃ gr:EmployeeGroup.T;
          (gr ∈ SELF.set ∧ e ∈ gr.set)
        => RAISEVAL = AlreadyEmployee
   METHOD T.getGroup(e)
     ENSURES e ∈ RESULT.set ∧ RESULT ∈ SELF.set
     EXCEPT ∀ gr:EmployeeGroup.T;
           ¬(gr ∈ SELF.set ∧ e ∈ gr.set)
        => RAISEVAL = NotEmployee
   METHOD T.createGroup(man)
     MODIFIES SELF.set
     ENSURES RESULT.manager = man
        ∧ RESULT.set' = {man}
        ∧ FRESH(RESULT)
        ∧ SELF.set' = insert(RESULT, SELF.set)
   METHOD T.removeGroup(g)
     MODIFIES SELF.set
     ENSURES SELF.set' = delete(g, SELF.set)
*>
END EmployeeDatabase.
```

FIGURE 6.6. EmployeeDatabase.i3, part 2

```
INTERFACE EmployeeSetFriend;
  IMPORT Employee, EmployeeSet, List;

  REVEAL EmployeeSet.T <: EmployeeSet.Public
    OBJECT
      cont: List.T
    END;

  PROCEDURE Sort (s: EmployeeSet.T);
< *
  TYPE_INVARIANT s: EmployeeSet.T
    size(s.set) = length(s.cont.l)
      ∧ ∀ e:Employee.T;
        e ∈ s.set ⇔ ∃ i:Int;
          0 ≤ i ∧ i < size(s.set)
            ∧ s.cont.l[i] = e

    STRENGTHEN EmployeeSet.T.insert(e)
    ENSURES
      e ∉ SELF.set
        ⇒ SELF.cont.l'[size(SELF.set)] = e

    PROCEDURE Sort(s)
    MODIFIES s.cont.l
    ENSURES
      ∀ i:Int;
        (0 ≤ i ∧ i < (size(s.set)-1))
          ⇒ (s.cont.l'[i]).ssnum
            ≤ (s.cont.l'[i+1]).ssnum
* >

END EmployeeSetFriend.
```

FIGURE 6.7. EmployeeSetFriend.i3

As before, the actual representation of a T is hidden, so we provide a specification field set, of abstract type EGSet.

EMPLOYEESETFRIEND

The type EmployeeSet.T, Figure 6.7, illustrates a Modula-3 *partial revelation* of an opaque type. It allows clients to know some of the detail of an EmployeeSet.T without exposing all of it.

In Figure 6.7, we expose the fact that an EmployeeSet.T has a field that is a List.T. We do not show the specification of the List interface here, but it has a specification field l that represents an abstract list. The TYPE_INVARIANT provides the abstraction relation, by relating this concrete field to the abstract fields visible from EmployeeSet. We strengthen the insert method specification in a consistent way, requiring that each new element be added at the end of the list.

Finally, we specify a procedure, Sort, that only makes sense in the presence of the revelation: the set abstraction does not have an order, but the list representation does. Since the modifies clause doesn't allow Sort to modify s.set, the specification can dispense with the usual clause saying that the final value must be a permutation of the initial value.

For more extensive use of partial revelation, see Chapter 6 of [69].

Chapter 7

Using LP to Debug LSL Specifications

In earlier chapters, we have attempted to show how Larch can be used to write precise specifications. However, it is not sufficient for specifications to be precise; they should also accurately reflect the specifier's intentions. Mistakes from many sources will crop up in specifications. Any practical methodology that relies on specifications must provide means for detecting and correcting their flaws, in short, for debugging them.

Parsing and type-checking are useful and easy to do, but don't go far enough. Unfortunately, we cannot *prove* the "correctness" of a specification, because there is no absolute standard against which to judge correctness. So we seek methods and tools that will be helpful in detecting and localizing the kinds of errors that we commonly observe.

Since the Larch style of specification emphasizes brevity and clarity rather than executability, it is usually not possible to evaluate Larch specifications by testing. Instead, LSL allows specifiers to state precise claims about specifications. If these claims are true, they can be verified statically. Such a verification won't guarantee that a specification meets a specifier's intent, but it is a powerful debugging technique. Once the flaws verification reveals are removed, there should be fewer doubts about the specification's accuracy.

The claims allowed in LSL specifications are undecidable in the general case. Hence we can't hope to build a tool that will automatically certify an arbitrary specification. However, tools can assist specifiers in checking claims during debugging.

This chapter describes how two such tools fit into our work on LSL. Our principal debugging tool is LP [30], the Larch proof assistant.[1] LP's design and development have been motivated primarily by our work on LSL, but it also has other uses (cf. Appendix E). Because of these other uses, and because we also intend to use LP to analyze Larch interface specifications, we have tried not to make LP too LSL-specific. Instead, we have chosen to build and use a second tool, the LSL Checker, as a front-end to LP. The LSL Checker checks the syntax and type consistency of LSL

---

[1] The version of LP described in this book is that released in November, 1991. A version with increased logical power is currently under development.

specifications, then generates LP proof obligations from their claims.

Sections 7.1 and 7.2 describe the checkable claims that can be made in LSL specifications. Sections 7.3 through 7.6 describe how LP is used to check these claims. Section 7.7 contains an extended example.

7.1   Semantic checks in LSL

We begin by reviewing the kinds of semantic claims that can be made in LSL. As mentioned in Chapter 4, semantic claims about LSL traits fall into three categories:

- consistency (that a specification does not contradict itself),

- theory containment (that a specification has intended consequences), and

- relative completeness (that a set of operators is adequately defined).

Consistency is an assertion about what is not in the theory of trait, and is therefore not expressible in LSL. Instead, it is implicitly required of all traits: no legal LSL trait's theory contains the inconsistent equation `true == false`. Claims in the other two categories are stated explicitly using the LSL constructs `implies` and `assumes`.

CHECKING IMPLICATIONS

An implies clause adds nothing to the theory of a trait. Instead, it makes a claim about theory containment. It enables specifiers to include information they believe to be redundant, either as a check on their understanding or to call attention to something that a reader might otherwise miss. The redundant information is of two kinds: statements like those in asserts clauses, which are claimed to be in the theory of the trait, and converts clauses, which describe the extent to which a specification is claimed to be complete.

The initial design of LSL incorporated a built-in notion of completeness. We quickly concluded, however, that requirements of completeness are better left to the specifier's discretion. It useful to check certain aspects of completeness long before a specification is finished. Furthermore, most finished specifications are left intentionally incomplete in places. LSL allows specifiers to make checkable claims about how complete they

```
LinearContainer(E, C): trait
  introduces
    empty: → C
    insert: E, C → C
    head: C → E
    tail: C → C
    isEmpty: C → Bool
    __ ∈ __: E, C → Bool
  asserts
    C generated by empty, insert
    C partitioned by head, tail, isEmpty
    ∀ c: C, e, e1: E
      head(insert(e, empty)) == e;
      tail(insert(e, empty)) == empty;
      isEmpty(empty);
      ¬isEmpty(insert(e, c));
      ¬(e ∈ empty);
      e ∈ insert(e1, c) == e = e1 ∨ e ∈ c
  implies
    ∀ c: C, e: E
      isEmpty(c) ⇒ ¬(e ∈ c)
    converts ∈, isEmpty
```

FIGURE 7.1. Sample LSL specification

intend specifications to be. These claims are usually most valuable during specification maintenance. Specifiers don't usually make erroneous claims about completeness when first writing a specification. On the other hand, when editing a specification, they often delete or change something without realizing its impact on completeness.

The first part of the implies clause of the trait LinearContainer,[2] Figure 7.1, asserts that if isEmpty of a container is true, no element is in that container. By checking that this assertion follows from the axioms of the trait, we can gain confidence that the axioms describing isEmpty and ∈ are appropriate.

---

[2]This trait is similar to the trait Container that appears in Figure 4.13 and in Appendix A: its theory is contained in that of Container. Many of the traits in this chapter are adapted from traits appearing in Appendix A. However, in order to better illustrate how traits are checked, we have changed them in small ways. In particular, we have often added implications and suppressed details that do not affect the points we wish to make.

```
PQ(E, Q): trait
  assumes TotOrd(E)
  includes LinearContainer(E, Q)
  asserts ∀ q: Q, e: E
    head(insert(e, q)) ==
      if isEmpty(q) then e
      else if e < head(q) then e
      else head(q);
    tail(insert(e, q)) ==
      if isEmpty(q) then empty
      else if e < head(q) then q
      else insert(e, tail(q))
  implies
    ∀ q: Q, e: E
    e ∈ q ⇒ ¬(e < head(q))
    converts isEmpty, head, tail, ∈
      exempting head(empty), tail(empty)
```

FIGURE 7.2. LSL specification for a priority queue

The converts clause in `LinearContainer` claims that the trait contains enough axioms to define ∈ and `isEmpty`; that is, given any fixed interpretations for the other operators, all interpretations of ∈ and `isEmpty` that satisfy the trait's axioms are the same.

The converts clause in `PQ`, Figure 7.2, involves more subtle checking. The exempting clause indicates that the lack of equations for `head(empty)` and `tail(empty)` is intentional: the operators `head` and `tail` are only claimed to be defined uniquely relative to interpretations for the terms `head(empty)` and `tail(empty)`. Section 7.5 describes the checking entailed by the converts clause in more detail.

CHECKING ASSUMPTIONS

There are two mechanisms for combining LSL specifications. Both are defined as operations on the texts of specifications. For both, the theory of a combined specification is axiomatized by the union of the axiomatizations for the individual specifications; each operator is constrained by the axioms of all traits in which it appears. Trait inclusion and trait assumption differ only in the checking they entail.

The trait `PQ`, Figure 7.2, which includes `LinearContainer`, further constrains the interpretations of `head`, `tail`, and `insert`. The assumes

```
TotOrd(E): trait
  introduces
    __ < __: E, E → Bool
    __ > __: E, E → Bool
  asserts forall x, y, z: E
    ¬( x < x );
    (x < y ∧ y < z) ⇒ x < z;
    x < y ∨ x = y ∨ y < x;
    x > y == y < x
  implies
    TotOrd(E, > for <, < for >)
    ∀ x, y: E
      ¬(x < y ∧ y < x)
```

FIGURE 7.3. LSL specification for total orders

clause of PQ indicates that PQ's theory also contains the theory of the trait TotOrd, Figure 7.3.

The use of `assumes` rather than `includes` entails additional checking. The assumption must be discharged whenever PQ is incorporated into another trait. For example, checking the trait

```
NumericPQ: trait
  includes PQ(N, NumericQ), Numeric
```

involves checking that the assertions in the trait TotOrd(N) are implied by those in the traits PQ, LinearContainer, and Numeric taken together. Sometimes these assumptions can be syntactically discharged for example, if Numeric explicitly includes TotOrd(N).

Figure 7.4 summarizes the checking that LSL requires for the sample traits introduced in this section.

7.2   Proof obligations for LSL specifications

An LSL specification generally consists of a hierarchy of traits, some of which include, assume, or imply others. We use the LSL Checker to syntax-check and type-check the traits, to extract the proof obligations required to check the semantic claims in the traits, and to discharge some of these proof obligations. This section describes how the LSL Checker extracts the proof obligations. The next several sections describe how we use LP to

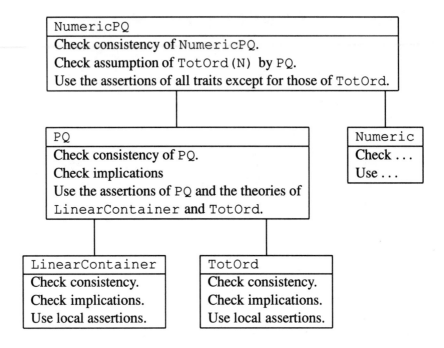

FIGURE 7.4. Summary of required checking

discharge those proof obligations that the LSL Checker cannot discharge syntactically.

To extract proof obligations, the LSL Checker computes the following sets of propositions (equations, generated by clauses, and partitioned by clauses) for each trait T in a trait hierarchy.

- The *assertions* of T consist of the propositions in the asserts clauses of T and of all traits (transitively) included in T.

- The *assumptions* of T consist of the assertions of all traits (transitively) assumed by T.

- The *axioms* of T consist of its assertions and its assumptions.

- The *immediate consequences* of T consist of the propositions in T's implies clause and the axioms of all traits that T explicitly implies.

The LSL Checker places the axioms for each trait T in a file named T_Axioms.lp. It also generates a file named T_Checks.lp, which contains the proof obligations associated with showing that T's axioms entail its immediate consequences, its converts clauses, and the assumptions of each trait explicitly included in or assumed by T. The LSL Checker does not generate an explicit proof obligation for showing that T's axioms are consistent. In fact, such a proof obligation is not expressible in LP. Like LSL, LP contains no mechanisms for making statements about what is not in a theory.

The LSL Checker can discharge some proof obligations syntactically, for example, because a proposition to be proved occurs textually among the axioms available for use in the proof. When it cannot do this, it places commands in T_Checks.lp that initiate a proof of the proposition. Sometimes LP will be able to carry out the required proof automatically; sometimes it will require user assistance.

Consider the trait NumericPQ, which includes both PQ and Numeric. Because PQ assumes TotOrd, it is necessary to check that the axioms of NumericPQ imply those of TotOrd. If Numeric explicitly includes or implies TotOrd, or if the assertions of TotOrd are among the axioms of Numeric, then the LSL Checker can discharge the assumption required for including PQ in NumericPQ. On the other hand, if Numeric simply asserts some properties of the binary relations < and >, the LSL Checker will formulate LP commands that initiate a proof of the conjecture that these properties imply the assertions of TotOrd.

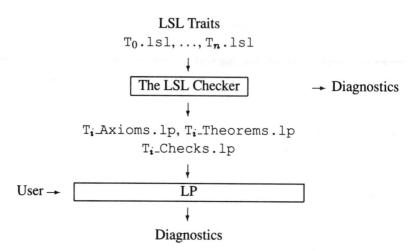

FIGURE 7.5. Using the LSL checker and LP to check LSL traits

LEMMAS FOR PROOF OBLIGATIONS

When checking the semantic claims in a hierarchy of traits, it is generally desirable to use lemmas that have been (or can be) shown separately to follow from the axioms of those traits. The *theorems* of a trait T consist of its axioms supplemented by all appropriately renamed propositions (transitively) implied by T or by some trait below T in the inclusion/assumption hierarchy.[3] The LSL Checker places the theorems for each trait T in a file named T_Theorems.lp, and refers to this file instead of T_Axioms.lp in T_Checks.lp when it is sound to do so. In general, soundness is guaranteed as long as there is a partial order for checking proof obligations in which each theorem is (or can be) checked before it is used as a lemma to discharge another proof obligation.

By providing a small set of axioms for a trait T, a specifier can make it easier to check traits that imply T or that include a trait that assumes T. By providing a large set of implications for T, a specifier can make it easier to reason about T and, in particular, to check traits that include or assume T, without at the same time making it harder to check traits that imply T or that include a trait that assumes T.

Figure 7.5 shows how the LSL Checker and LP are used together to check LSL traits.

---

[3]Some generated by and partitioned by clauses will not qualify as theorems of T when a renaming identifies the generated or partitioned sort with some other sort.

```
declare sorts
  C, E
  ..
declare operators
  head: C → E
  insert: E, C → C
  isEmpty: C → Bool
  tail: C → C
  empty: → C
  ∈: E, C → Bool
  ..
declare variables
  e: E
  c: C
  e1: E
  ..
```

FIGURE 7.6. LP declarations produced from `LinearContainer`

7.3   Translating LSL traits into LP

LP is a proof assistant for a subset of multisorted first-order logic with
equality. The basis for proofs in LP is called a *logical system*. This section
contains an overview of the components of a logical system in LP and
discusses their relation to the components of an LSL trait. The following
sections discuss how these components are used by LP to discharge proof
obligations associated with LSL traits.

A logical system in LP consists of a signature (given by declarations)
plus equations, rewrite rules, operator theories, induction rules, and
deduction rules. Logical systems are closely related to LSL theories, but
are handled somewhat differently. Axioms in LP have operational as well
as semantic content, and they can be presented to LP incrementally, rather
than all at once.

DECLARATIONS

Sorts, operators, and variables play the same roles in LP as they do in LSL.
As in LSL, operators and variables must be declared, and operators can be
overloaded. There are a few minor differences: sorts must be declared in
LP, and LP doesn't provide scoping for variables.

The LSL Checker produces the declarations in Figure 7.6 from the introduces and ∀ clauses in the trait `LinearContainer`.

EQUATIONS AND REWRITE RULES

Equations play a prominent role in LP. Some of LP's inference mechanisms work directly with equations. Most, however, require that equations be oriented into *rewrite rules*, which LP uses to reduce terms to normal forms. It is usually essential that the rewriting relation be *terminating*, that is, no term can be rewritten infinitely many times. LP provides several mechanisms that automatically orient many sets of equations into terminating rewriting systems. For example, in response to the commands

```
set name group
declare sort G
declare variables x, y, z: G
declare operators e: → G, i: G → G, *: G, G → G
assert
  (x*y)*z == x*(y*z)
  e == i(x)*x
  e*x == x
  ..
```

which enter the usual axioms for groups, LP produces the rewrite rules

```
group.1: (x * y) * z → x * (y * z)
group.2: i(x) * x → e
group.3: e * x → x
```

LP automatically reverses the second equation to prevent nonterminating rewriting sequences such as

```
e → i(e) * e → i(e) * i(e) * e → ...
```

A system's *rewriting theory* consists of the propositions that can be proved by reduction to normal form. This theory is always a subset of its *equational theory*, which consists of the propositions that can be proved from its equations and from its rewrite rules considered as equations. A system's rewriting theory does not usually include all of its equational theory. The proof mechanisms discussed in Section 7.4 help to compensate for this incompleteness. In the case of group theory, for example, the equation `e == i(e)` follows logically from the axioms, but is not in the rewriting theory of the three rewrite rules: it is irreducible, but not an identity.

LP provides built-in rewrite rules to simplify predicates involving the connectives ¬, ∧, ∨, ⇒, and ⇔, the equality operator =, and the conditional operator `if`. These rewrite rules are sufficient to prove many identities involving these operators, but not all. Unfortunately, the sets of rewrite rules that are known to be complete for propositional calculus require exponential time and space. Furthermore, they tend to expand, rather than simplify, propositions that do not reduce to identities. These are serious drawbacks when we are debugging specifications, because we often attempt to prove conjectures that are not true. So none of the complete sets of rewrite rules is built into LP. Instead, LP provides proof mechanisms that can be used to overcome incompleteness in a rewriting system. It also allows users to add any of the complete sets they choose to use.

LP treats the equations `true == false` and `x = t == false`, where `t` is a term not containing the variable `x`, as inconsistent. (The second equation rules out empty sorts.) Inconsistencies can be used to establish subgoals in proofs by cases and contradiction. If they arise in other situations, they indicate that the axioms in the logical system are inconsistent.

OPERATOR THEORIES

LP provides special mechanisms for handling some equations that cannot be oriented into terminating rewrite rules. LP recognizes two operator theories: the commutative theory and the associative-commutative (ac) theory. For example, the command `assert ac +` says that `+` is associative and commutative. Logically, this assertion is an abbreviation for two equations:

```
x + (y + z) == (x + y) + z
x + y == y + x
```

Operationally, it causes LP to match and unify terms modulo associativity and commutativity. This increases the number of theories that LP can reason about. It also reduces the number of axioms required to describe various theories, the number of reductions necessary to derive identities, and the need for certain kinds of user interaction, such as case analysis. Its main drawback is that it can be much slower than ordinary rewriting.[4]

---

[4]A secondary drawback is that ordering equations that contain commutative and ac operators into terminating sets of rewrite rules is, in principle, more difficult. In practice, however, this is not a problem.

INDUCTION RULES

LP uses induction rules to generate subgoals in proofs by induction. The syntax for induction rules is the same in LP as in LSL.[5]

Users can specify multiple induction rules for a single sort and can use the appropriate rule when attempting to prove an equation by induction. For example, assuming appropriate declarations, the LP commands

```
set name setInduction1
assert S generated by empty, insert
set name setInduction2
assert S generated by empty, singleton, ∪
```

allow

```
prove x ⊆ x by induction using setInduction2
```

In LSL, the axioms of a trait typically have only one generated by for a sort. It is often useful, however, to put others in the trait's implications.

DEDUCTION RULES

LP subsumes the logical power of the partitioned by construct of LSL in deduction rules, which LP uses to deduce equations from other equations and rewrite rules. Like other formulas in LP, deduction rules may be asserted as axioms or proved as theorems. While the partitioned by clause in the trait LinearContainer can be expressed by an equation, in general a partitioned by clause is equivalent to a universal-existential axiom, which can only be expressed as a deduction rule in LP. For example, the LP commands

```
assert S partitioned by ∈
assert
  when (∀ e) e ∈ x == e ∈ y
  yield x == y
```

are equivalent and define a deduction rule equivalent to the axiom of set extensionality

$$(\forall x, y : S) \left[(\forall e : E)(e \in x \Leftrightarrow e \in y) \Rightarrow x = y \right]$$

This deduction rule enables LP to deduce equations such as x == x ∪ x automatically from equations such as e ∈ x == e ∈ (x ∪ x).

---

[5]The semantics of induction is somewhat stronger in LSL than in LP, since arbitrary first-order formulas cannot be written in this version of LP.

Deduction rules can have multiple hypotheses and/or multiple conclusions. For example, the transitivity of $<$ can be formulated as a deduction rule with two hypotheses:

```
when i < j, j < k yield i < k
```

The built-in $\wedge$-splitting law is a deduction rule with two conclusions:

```
when p ∧ q yield p, q
```

Such deduction rules serve to improve the performance of LP and to reduce the need for user interaction.

LP automatically applies deduction rules to equations and rewrite rules whenever they are normalized. The sample proof in Section 7.5 illustrates the logical power of deduction rules, as well as the benefits of applying them automatically to the case and induction hypotheses in a proof.

7.4   Proof mechanisms in LP

This section provides a brief overview of the proof mechanisms in LP. The next two sections discuss how they are used to check LSL semantic claims.

LP provides mechanisms for proving theorems using both forward and backward inference. Forward inferences produce consequences from a logical system; backward inferences produce subgoals whose proof will suffice to establish a conjecture. There are four methods of forward inference in LP.

1. Automatic *normalization* produces new consequences when a rewrite rule is added to a system. LP keeps rewrite rules, equations, and deduction rules in normal form.

   If an equation or rewrite rule normalizes to an identity, it is discarded, because it is logically and operationally superfluous. If all hypotheses of a deduction rule normalize to identities, the deduction rule is replaced by the equations in its conclusions. If all conclusions of a deduction rule normalize to identities, the deduction rule is discarded.

   Users can "immunize" equations, rewrite rules, and deduction rules to protect them from automatic normalization, both to enhance the performance of LP and to preserve a particular form for use in a proof. Users can also "deactivate" rewrite rules and deduction rules to prevent them from being applied automatically.

2. Automatic *application of deduction rules* produces new consequences after equations and rewrite rules in a system are normalized. Deduction rules can also be applied by explicit command, for example, to immune equations.

3. The computation of *critical-pair equations* and the Knuth-Bendix *completion procedure* [58, 72] produce equational consequences (such as i(e) == e) from incomplete rewriting systems (such as the three rewrite rules for groups, page 130). We often compute critical-pair equations from selected sets of rewrite rules. Sometimes we run the completion procedure to find the last few consequences to finish off a proof or, as discussed in Section 7.7, to look for inconsistencies. However, we rarely complete our rewriting systems, because a complete set of rewrite rules with a given equational theory may not exist, may be too expensive to obtain, or may lead to normal forms that are hard to read [28].

4. Explicit *instantiation* of variables in equations, rewrite rules, and deduction rules also produces consequences. For example, in a system that contains the deduction rule

   when (∀ e) e ∈ x == e ∈ y yield x == y

   and the rewrite rule e ∈ (x ∪ y) → e ∈ x ∨ e ∈ y, we can instantiate y in the deduction rule by x ∪ x to produce the conclusion x == x ∪ x.

There are seven methods of backward inference for proving theorems in LP. These methods are invoked by the prove and resume commands. In each method, LP generates a set of subgoals to be proved, that is, lemmas that together are sufficient to imply the conjecture. For some methods, LP generates additional hypotheses that may be used to prove particular subgoals.

1. *Normalization* rewrites conjectures. If a conjecture normalizes to an identity, it is a theorem. Otherwise the normalized conjecture becomes the subgoal to be proved.

2. *Proofs by cases* can further normalize a conjecture. The command prove e by cases $t_1, \ldots, t_n$, where $t_1, \ldots, t_n$ are predicates, directs LP to prove an equation e by division into cases

$t_1, \ldots, t_n$ (or into two cases, t_1 and $\neg t_1$, if $n = 1$). When $n > 1$, one subgoal is to prove that the cases are exhaustive, i.e., $t_1 \lor \ldots \lor t_n$. In addition, for each case t_i, LP substitutes new constants for the variables of t_i in both t_i and e to form t_i' and e_i', which it uses to creates the subgoal e_i' with the additional hypothesis $t_i' \to$ true. If an inconsistency results from adding the case hypothesis t_i', that case is impossible, and e_i' is vacuously true. Otherwise, the subgoal e_i' must be shown to follow from the axioms supplemented by the case hypothesis.

Case analysis has two primary uses. If the conjecture is a theorem, a proof by cases may circumvent a lack of completeness in the rewrite rules. If the conjecture is not a theorem, an attempted proof by cases may simplify the conjecture and make it easier to understand why the proof is not succeeding.

3. *Proofs by induction* are based on the induction rules described in Section 7.3. For example, a proof by induction of

```
isEmpty(c)  ⇒  ¬(e ∈ c)
```

from the axioms of `LinearContainer` involves two steps. The *basis step* involves showing that

```
isEmpty(empty)  ⇒  ¬(e ∈ empty)
```

This follows from the axioms by normalization. The *induction step* involves picking a new constant cc, assuming

```
isEmpty(cc)  ⇒  ¬(e ∈ cc)
```

as an *induction hypothesis*, and showing that

```
isEmpty(insert(e1, cc))  ⇒
      ¬(e ∈ insert(e1, cc))
```

This follows by normalization from the axioms supplemented by this induction hypothesis.

4. *Proofs by contradiction* provide an indirect method of proof. If an inconsistency follows from adding the negation of the conjecture to LP's logical system, then the conjecture is a theorem.

5. *Proofs of implications* can be carried out using a simplified form of proof by cases. The command prove $t_1 \Rightarrow t_2$ by $\Rightarrow$ directs LP to prove the subgoal t_2' using the hypothesis $t_1' \rightarrow$ true, where t_1' and t_2' are obtained as in a proof by cases. This suffices because the implication is vacuously true when t_1' is false.

6. *Proofs of conditionals* can also be carried out using a simplified form of proof by cases. The command

   prove if(t_1, t_2, t_3) == t_4 by if

   directs LP to prove the subgoal t_2' == t_4' using the hypothesis t_1', and to prove the subgoal t_3' == t_4' using the hypothesis $\neg t_1'$, where $t_1', \ldots, t_4'$ are obtained as in a proof by cases.

7. *Proofs of conjunctions* provide a way to reduce the expense of rewriting modulo the associativity and commutativity of $\wedge$. The command prove $t_1 \wedge \ldots \wedge t_n$ by $\wedge$ directs LP to prove each of $t_1, \ldots, t_n$ as a separate subgoal.

LP allows users to specify which methods of backward inference are applied automatically and in what order. This is done by using the set proof-methods command. For example, the LP command

   set proof-methods if, $\Rightarrow$, normalization

tells LP that whenever it is given a conjecture to prove, it should automatically try to apply these three methods, in the given order.

LP also provides automatic methods of backward inference for proving induction and deduction rules. In each method, LP generates a set of subgoals to be proved, as well as additional hypotheses that may be used to prove particular subgoals. (See the next section for examples.)

Proofs of interesting conjectures hardly ever succeed on the first try. Sometimes the conjecture is wrong. Sometimes the formalization is incorrect or incomplete. Sometimes the proof strategy is flawed or not detailed enough. When an attempted proof fails, we use a variety of LP facilities (e.g., case analysis) to try to understand the problem. Because many proof attempts fail, LP is designed to fail relatively quickly and to provide useful information when it does. It is not designed to find difficult proofs automatically. Unlike the Boyer-Moore prover [8], it does not perform heuristic searches for a proof. Unlike LCF [71], it does not allow users to define complicated search tactics. Strategic decisions, such as when to try induction, must be supplied as explicit LP commands.

```
declare sorts
  E
  ‥
declare operators
  < : E, E → Bool
  > : E, E → Bool
  ‥
declare variables
  x: E
  y: E
  z: E
  ‥
set name TotOrd
assert
  ¬ (x < x)
  (x < y ∧ y < z) ⇒ x < z
  x < y ∨ x = y ∨ y < x
  x > y == y < x
  ‥
```

FIGURE 7.7. TotOrd_Axioms.lp

On the other hand, LP is more than a "proof checker," since it does not require proofs to be described in minute detail. In many respects, LP is best described as a "proof debugger."

7.5   Checking theory containment

The proof obligations triggered by implies and assumes clauses in an LSL trait require us to check theory containment, that is to check that claims follow from axioms. This section discusses how the LSL Checker formulates the proof obligations for theory containment for LP, as well as how we use LP to discharge these obligations. The next section discusses checking consistency.

PROVING AN EQUATION

The proof obligation for an equation is easy to formulate. Consider, for example, the proof obligations that must be discharged to check the trait TotOrd shown in Figure 7.3. Figure 7.7 displays the LP commands that the LSL Checker extracts from this trait in order to axiomatize

```
execute TotOrd_Axioms
set name TotOrdTheorem
% Prove implication of TotOrd(E, > for <, < for >)
prove ¬(x > x)
   qed
prove (x > y ∧ y > z) ⇒ x > z
   qed
prove x > y ∨ x = y ∨ y > x
   qed
prove x < y == y > x
   qed
% Prove implied equation
prove ¬(x < y ∧ y < x)
   qed
```

FIGURE 7.8. TotOrd_Checks.lp

its theory, and Figure 7.8 displays the LP commands that the LSL Checker extracts from this trait in order to discharge its proof obligations. The execute command obtains the axioms for TotOrd from the file TotOrd_Axioms.lp. The prove commands initiate proofs of the five immediate consequences of TotOrd.

LP can discharge all proof obligations except the first without user assistance. The user is alerted to the need to supply assistance in this proof by a diagnostic ("Proof suspended") printed in response to the qed command. At this point, the user can complete the proof by entering the complete command or the command

```
critical-pairs TotOrd with TotOrd
```

Proofs of equations require varying amounts of assistance. For example, when checking that LinearContainer implies

```
isEmpty(c) ⇒ ¬(e ∈ c)
```

the single LP command resume by induction suffices to finish the proof.

When checking that PQ, Figure 7.2, implies

```
e ∈ q ⇒ ¬(e < head(q))
```

more guidance is required. This proof proceeds by induction on q. LP proves the basis subgoal without assistance. For the induction subgoal, LP introduces a new constant qc to take the place of the universally-quantified variable q, adds

```
e ∈ qc ⇒ ¬ (e < head(qc))
```

as the induction hypothesis, and attempts to prove

```
e ∈ insert(e1, qc) ⇒
    ¬ (e < head(insert(e1, qc)))
```

which normalizes to

```
(e1 = e ∨ e ∈ qc) ⇒
    ¬ (e < (if isEmpty(qc) then e1
            else if e1 < head(qc) then e1
            else head(qc)))
```

LP now automatically applies the ⇒ proof method, i.e., it assumes the hypothesis of the implication, introducing new constants ec and e1c to take the place of the variables e and e1, and attempts to prove the conclusion of the implication from this hypothesis. At this point, further guidance is required. The command

```
resume by case isEmpty(qc)
```

directs LP to divide the proof into two cases based on the predicate in the first if. In the first case, isEmpty(qc), the desired conclusion normalizes to ¬ (ec < e1c). The complete command leads LP to deduce ¬ (e ∈ qc), using the implied equation in the trait LinearContainer, which is available for use in the proof because LinearContainer precedes PQ in the trait hierarchy. With this fact, LP is able to finish the proof in the first case automatically. The second case, ¬isEmpty(qc), requires more user assistance.

Figure 7.9 shows the entire proof, as recorded and annotated by LP in a *script file*. In addition to recording user input, LP has indented the script to reveal the structure of the proof, and it has annotated the proof by adding lines (beginning with <>) to indicate when it introduced subgoals and lines (beginning with [ ]) to indicate when each of these subgoals and the theorem itself were proved. Such an annotated proof provides the user with a means of regression testing after changing the axioms for a trait. On request, when LP executes the annotated proof (using the new set of axioms), it will halt execution and print an error message if the annotations do not match the execution. These checks help pinpoint the source of a problem when changes in the axioms cause some step in the proof to succeed with less user guidance than expected or to require more guidance. Without the check, LP might, for example, apply a tactic intended for a

```
set proof-methods ⇒, normalization
prove e ∈ q ⇒ ¬(e < head(q)) by induction
   <> 2 subgoals for proof by induction on 'q'
     <> 1 subgoal for proof of ⇒
       [] ⇒ subgoal
     [] basis subgoal
     <> 1 subgoal for proof of ⇒
       resume by case isEmpty(qc)
       <> 2 subgoals for proof by cases
         % Handle case isEmpty(qc)
         complete
         [] case isEmpty(qc)
         % Handle case ¬isEmpty(qc)
         resume by case e1c < head(qc)
         <> 2 subgoals for proof by cases
           % Handle case e1c < head(qc)
           resume by contradiction
           <> 1 subgoal for proof by contradiction
             complete
             [] contradiction subgoal
           [] case e1c < head(qc)
           % Handle case ¬(e1c < head(qc))
           resume by contradiction
           <> 1 subgoal for proof by contradiction
             complete
             [] contradiction subgoal
           [] case ¬(e1c < head(qc))
         [] case ¬(isEmpty(qc))
       [] ⇒ subgoal
     [] induction subgoal
   [] conjecture
qed
```

FIGURE 7.9. LP-annotated proof of PQ implication

```
FinSet: trait
  introduces
    empty:  → S
    insert: S, E → S
    singleton: E → S
    __ ∪ __ : S, S → S
    __ ∈ __ : E, S → Bool
    __ ⊆ __ : S, S → Bool
  asserts
    S generated by empty, insert
    S partitioned by ∈
    forall s, s1: S, e, e1: E
      singleton(e) == insert(empty, e);
      s ∪ empty == s;
      s ∪ insert(s1, e) == insert(s ∪ s1, e);
      ¬ (e ∈ empty);
      e ∈ insert(s, e1) == e ∈ s ∨ e = e1;
      empty ⊆ s;
      insert(s, e) ⊆ s1 == s ⊆ s1 ∧ e ∈ s1
  implies
    S partitioned by ⊆
    S generated by empty, singleton, ∪
```

FIGURE 7.10. An LSL trait for finite sets

particular case in a proof to the wrong case, thereby causing the proof to fail in mysterious ways. This checking helps prevent proofs from getting "out of sync" with their author's conception of how they should proceed.

PROVING A "PARTITIONED BY"

Proving a partitioned by clause amounts to proving the validity of the associated deduction rule in LP. For example, the LSL Checker formulates the proof obligations associated with the partitioned by in the implies clause of Figure 7.10 using the LP commands

```
execute FinSet_Axioms
prove S partitioned by ⊆
```

and LP translates the partitioned by into the deduction rule

```
when (∀ s3) s1 ⊆ s3 == s2 ⊆ s3,
              s3 ⊆ s1 == s3 ⊆ s2
  yield s1 == s2
```

LP initiates a proof of this deduction rule by introducing two new constants, `s1c` and `s2c` of sort S, assuming `s1c ⊆ s3 == s2c ⊆ s3` and `s3 ⊆ s1c == s3 ⊆ s2c` as additional hypotheses, and attempting to prove the subgoal `s1c == s2c`. LP cannot prove `s1c == s2c` directly, because the equation is irreducible. The user can guide LP by typing `complete`, which yields the lemma `e ∈ s1c == e ∈ s2c`, after which LP automatically finishes the proof by applying the deduction rule associated with the assertion `S partitioned by ∈`.

PROVING A "GENERATED BY"

Proving a generated by clause involves proving that the set of elements generated by the given operators contains all elements of the sort. For example, the LSL Checker formulates the proof obligation associated with the generated by in the implies clause of Figure 7.10 as

```
execute FinSet_Axioms
prove S generated by empty, singleton, ∪
```

LP then introduces a new operator `isGenerated:S→Bool`, adds the hypotheses

```
isGenerated(empty)
isGenerated(singleton(e))
(isGenerated(s1) ∧ isGenerated(s))
    ⇒ isGenerated(s1 ∪ s)
```

and attempts to prove the subgoal `isGenerated(s)`. User guidance is required to complete the proof, for example, by entering the commands

```
resume by induction
complete
```

directing LP to use the induction scheme obtained from the assertion

```
S generated by empty, insert
```

and then to run the completion procedure to draw the necessary inferences from the additional hypotheses.

PROVING A "CONVERTS"

Proving that a trait converts a set of operators involves showing that the axioms of the trait define the operators in the set relative to the other operators in the trait. For example, to show that LinearContainer

```
execute LinearContainer_Theorems
declare operators
  isEmpty': C → Bool
  ∈': E, C → Bool
  ..
assert C partitioned by head, tail, isEmpty'
assert
  isEmpty'(empty)
  ¬(isEmpty'(insert(e, c)))
  ¬(e ∈' empty)
  e ∈' insert(e1, c) == e = e1 ∨ e ∈' c
  isEmpty'(c) ⇒ ¬(e ∈' c)
  ..
set name conversionChecks
prove e ∈ c == e ∈' c
  qed
prove isEmpty(c) == isEmpty'(c)
  qed
```

FIGURE 7.11. Proof obligations for converts in LinearContainer

converts isEmpty and ∈, one must show that, given any interpretations for empty and insert, there are unique interpretations for
isEmpty and ∈ that satisfy the axioms of LinearContainer.
Equivalently, we must show that the theories of LinearContainer and
LinearContainer (isEmpty' for isEmpty, ∈' for ∈) together imply the two equations isEmpty(c) == isEmpty'(c) and
e ∈ c == e ∈' c.

The LSL Checker formulates these proof obligations with the LP
commands in Figure 7.11.[6] The only user guidance required to discharge
these proof obligations is a command to proceed by induction.

The proof obligation for the converts clause in PQ is similar. Here we
must show that given any interpretations for empty and insert, as well
as for the exempted terms head(empty) and tail(empty), there
are unique interpretations for head, tail, isEmpty, and ∈ that satisfy
the theory of PQ. The proof obligations for this are shown in Figure 7.12.
Again, the only user guidance needed to complete the proofs are commands
to proceed by induction.

---

[6]The figure's last assertion comes from the implies clause in LinearContainer.

```
execute PQ_Theorems
% Declarations, axioms, and theorems for
%   head', tail', isEmpty', ∈' occur here
set name exemptions
assert
  head(empty) == head'(empty)
  tail(empty) == tail'(empty)
  ..
set name conversionChecks
prove isEmpty(q) == isEmpty'(q)
  qed
prove head(q) == head'(q)
  qed
prove tail(q) == tail'(q)
  qed
prove e ∈ q == e ∈' q
  qed
```

FIGURE 7.12. Proof obligations for converts in PQ

7.6   Checking consistency

Checks for theory containment fall into the typical pattern of use of a theorem prover. The check for consistency is harder to formulate because it involves nonconsequence rather than consequence. Techniques for detecting when this check fails are more useful than techniques for certifying that it succeeds.

A standard approach in logic to proving consistency involves interpreting the theory being checked in another theory whose consistency is assumed (e.g., Peano arithmetic) or has been established previously [77]. In this approach, user assistance is required to define the interpretation. The proof that the interpretation satisfies the axioms of the trait being checked then becomes a problem of showing theory containment, for which LP is well suited. This approach is cumbersome and unattractive in practice. More promising approaches are based on metatheorems in first-order logic that can be used for restricted classes of specifications. For example, any extension by definitions (see [77]) of a consistent theory is consistent.

For equational traits (i.e., traits with purely equational axiomatizations, of which there are relatively few), questions about consistency can be translated into questions about critical pairs. In some cases, we can use LP to answer these questions by running the completion procedure or by computing critical pairs. If these actions generate an inconsistency, the axioms are inconsistent; if they complete the axioms without generating the equation `true == false`, then the trait is consistent. This proof strategy will not usually succeed in proving consistency, because many equational theories cannot be completed at all, or cannot be completed in an acceptable amount of time and space. However, it has proved useful in finding inconsistencies among equations.

We can use all of LP's forward inference mechanisms to search for inconsistencies in a specification. The completion procedure searches for inconsistencies automatically, and we can instantiate axioms by "focus objects" (in the sense of McAllester [64]) to provide the completion procedure with a basis for its search. Even though unsuccessful searches do not certify that a specification is consistent, they increase our confidence in a specification, just as testing increases our confidence in a program.

```
Coordinate: trait
  introduces
    origin: → Coord
    __ - __: Coord, Coord → Coord
  asserts ∀ cd: Coord
    cd - cd == origin

Region(R): trait
  assumes Coordinate
  introduces
    __ ∈ __: Coord, R → Bool
    % cd ∈ r is true if point cd is in region r
    % Nothing is assumed about the contiguity
    %   or shape of regions

Displayable(T): trait
  assumes Coordinate
  includes Region(T)
  introduces
    __[__]: T, Coord → Color
    % t[cd] represents appearance of object t
    % at point cd
```

FIGURE 7.13. Prototype traits for windowing abstraction

7.7   Extended example

To illustrate our approach to checking specifications in a slightly more realistic setting, we show how one might construct and check some traits to be used in the specification of a simple windowing system [43]. These are preliminary versions of traits that would likely be expanded as the specifications (including the interface parts) were developed.

The first three traits, Figure 7.13, declare the signatures of some basic operators.

The proof obligations associated with these traits are easily discharged. When LP's completion procedure is run on Coordinate, it terminates without generating any critical pairs. Since Coordinate has no generated by or partitioned by clauses, this is sufficient to guarantee that it is consistent. When checking the inclusion of Region by Displayable, Region's assumption of Coordinate is discharged syntactically, using Displayable's assumption of the same trait.

```
Window(W): trait
  assumes Coordinate
  includes Region, Displayable(W)
  W tuple of cont, clip: R, fore, back: Color, id: WId
  asserts ∀ w: W, cd: Coord
    cd ∈ w == cd ∈ w.clip;
    w[cd] == if cd ∈ w.cont then w.fore else w.back
  implies converts __[__], ∈:Coord,W→Bool
```

FIGURE 7.14. Window.lsl

The Window trait, Figure 7.14, defines a window as an object composed of content and clipping regions, foreground and background colors, and a window identifier. The operator ∈ is qualified by a signature in the last line of the trait because it is overloaded, and it is necessary to say which ∈ is converted.

There are three proof obligations associated with this trait. The assumptions of Coordinate in Region and Displayable are syntactically discharged using Window's assumption. The converts clause is discharged by LP without user assistance. The other proof obligation is consistency. As discussed in the previous section, we use the completion procedure to search for inconsistencies. Running it for a short time neither uncovers an inconsistency nor proves consistency.

The View trait, Figure 7.15, defines a view as an object composed of windows at locations. There are several proof obligations associated with this trait. Once again, the assumptions of Window and Displayable are discharged syntactically by the assumption in View. Once again, using the completion procedure to search for inconsistencies uncovers no problems. However, checking the converts clause does turn up a problem. The conversion of inW and both ∈'s is easily proved by induction over objects of sort V. However, the inductive base case for __[__] does not reduce at all, because emptyV[cd] is not defined. This problem can be solved either by defining emptyV[cd] or by adding

```
exempting ∀ cd: Coord emptyV[cd]
```

to the converts clause. We choose the latter because there is no obvious definition for emptyV[cd]. With the added exemption, the inductive proof of the conversion of __[__] goes through without further interaction.

When we attempt to prove the first of the explicit equations in the implies clause of View, we run into difficulty. After automatically applying its

```
View: trait
  assumes Coordinate
  includes Window, Displayable(V)
  introduces
    emptyV: → V
    addW: V, Coord, W → V
    __ ∈ __: W, V → Bool
    inW: V, WId, Coord → Bool
  asserts
    V generated by emptyV, addW
    ∀ cd, cd1: Coord, v: V, w, w1: W, wid: WId
      ¬(cd ∈ emptyV);
      cd ∈ addW(v, cd1, w) ==
        (cd - cd1) ∈ w ∨ cd ∈ v;
      ¬(w ∈ emptyV);
      w ∈ addW(v, cd1, w1) == w.id = w1.id ∨ w ∈ v;
      addW(v, cd1, w)[cd] ==
        if (cd - cd1) ∈ w
          then w[cd - cd1] else v[cd];
      % In view only if in a window, offset by origin
      ¬inW(emptyV, wid, cd);
      inW(addW(v, cd, w), wid, cd1) ==
        (w.id = wid ∧ (cd - cd1) ∈ w)
          ∨ inW(v, wid, cd1)
  implies
    ∀ cd, cd1: Coord, v,v1: V, w: W
      % New window does not affect the appearance
      % of parts of the view lying outside the window
      ¬inW(addW(v, cd, w), w.id, cd1)
        ⇒ addW(v, cd, w)[cd1] = v[cd1];
      % Appearance within newly added window is
      % independent of the view to which it is added
      inW(addW(v, cd1, w), w.id, cd)
        ⇒ addW(v, cd1, w)[cd] = addW(v1, cd1, w)[cd]
  converts inW, ∈:Coord,V→Bool, ∈:W,V→Bool,
            __[__]:V,Coord→Color
```

FIGURE 7.15. Preliminary version of View.lsl

proof method for implications, LP reduces the conjecture to

```
if (cd1c - cdc) ∈ wc.clip
    then if (cd1c - cdc) ∈ wc.cont
            then wc.fore else wc.back
    else vc[cd1c]
== vc[cd1c]
```

and reduces the assumed hypothesis of the implication to

```
¬ ((cdc - cd1c) ∈ wc.clip)
```

At this point, we ask ourselves why the hypothesis is not sufficient to reduce the conjecture to an identity. The problem seems to be the order of the operands of −. This leads us to look carefully at the second equation for inW in trait View. There we discover that we have written cd − cd1 when we should have written cd1 − cd, or, to be consistent with the form of the other equations, reversed the role of cd and cd1 in the left side of the equation. After changing this axiom to

```
inW(addW(v, cd1, w), wid, cd) ==
    (w.id = wid ∧ (cd - cd1) ∈ w)
      ∨ inW(v, wid, cd)
```

the proof of the first implication goes through without interaction.

The second conjecture, after LP applies its proof method for implications, reduces to

```
if (cdc - cd1c) ∈ wc.clip
    then if (cdc - cd1c) ∈ wc.cont
            then wc.fore else wc.back
    else vc[cdc]
  ==
if (cd - cd1c) ∈ wc.clip
    then if (cdc - cd1c) ∈ wc.cont
            then wc.fore else wc.back
    else v'[cdc]
```

We resume the proof by dividing it into two cases based on the predicate in the outermost if's. When this predicate is true, the conjecture reduces to true; when it is false, the conjecture reduces to

```
vc[cdc] == v'[cdc]
```

Since v' is a variable and vc a new constant, we know that we are not going to be able to reduce this to true. This does not necessarily mean that the proof will fail, since we could be in an impossible case (i.e., the

current hypotheses could lead to a contradiction). However, examining the
current hypotheses,

```
inW(vc, wc.id, cdc)              % Hypothesis of ⇒
¬((cdc - cdlc) ∈ wc.clip)        % Case hypothesis
```

gives us no obvious reason to believe that a contradiction exists.

This leads us to wonder about the validity of the conjecture we are trying
to prove, and to ask ourselves why we thought it was true when we added
it to the trait. Our informal reasoning had been:

1. The hypothesis inW(addW(v, cdl, w), w.id, cd) of the
   conjecture guarantees that coordinate cd is in window w in the view
   addW(v, cdl, w).

2. If w is added at the same place in v' as in v, cd must also be in
   addW(v', cdl, w).

3. Furthermore cd - cdl will be the same relative coordinate in w
   in both addW(v, cdl, w) and addW(v', cdl, w).

4. Therefore the equation

```
addW(v, cdl, w)[cd] ==
    if (cd - cdl) ∈ w
        then w[cd -cdl] else v[cd]
```

   in trait View should guarantee the conclusion.

The first step in formalizing this informal argument is to attempt to prove

```
inW(addW(v, cdl, w), w.id, cd) ⇒ (cd - cdl) ∈ w
```

as a lemma. LP reduces the conclusion of this implication to

```
(cdc - cdlc) ∈ wc.clip
```

using the normalized implication hypothesis

```
(cdc - cdlc) ∈ wc.clip ∨ inW(vc, wc.id, cdc)
```

Casing on the first disjunct of the hypothesis reduces the conjecture to
false under the same implication and case hypotheses as above.

We are thus stuck in the same place as in our attempted proof of the
original conjecture. This leads us to question the validity of the first step
in our informal proof, and we discover a flaw there: when v contains a
window with the same id as w, the implication is not sound. The problem

is that we implicitly assumed the invariant that no view would contain two windows with the same id, and our specification does not guarantee this. There are several ways around this problem, among them:

1. Trait View could be changed so that addW chooses a unique id whenever a window is added.

2. Trait View could be changed so that addW is the identity function when the id of the window to be added is already associated with a window in the view.

3. The preservation of the invariant could be left to the interface level.

We choose the third alternative and weaken the second implication of trait View to:

```
∀ cd, cd1: Coord, v, v': V, w: W
  % Appearance within a newly added window is
  % independent of the view to which it is added,
  % provided that the window id is not already
  % present in the view.
  (¬ (w ∈ v) ∧ ¬ (w ∈ v')
        ∧ inW(addW(v, cd1, w), w.id, cd))
     ⇒ addW(v, cd1, w)[cd] = addW(v', cd1, w)[cd]
```

which is proved with a small amount of user interaction after proving the lemma

```
¬ (w ∈ v) ⇒ ¬inW(v, w.id, cd)
```

by induction on v.

Finally, we introduce a coordinate system.

```
CartesianView: trait
  includes View, Natural
  Coord tuple of x, y: N
  asserts ∀ cd, cd1: Coord
    origin == [0, 0];
    cd - cd1 == [cd.x ⊖ cd1.x, cd.y ⊖ cd1.y]
  implies converts origin, -
```

LP uses the facts of the trait Natural (see Appendix A) to automatically discharge the assumption of Coordinate that has been carried from level to level. LP requires no assistance to complete the proof that the coordinate operators are indeed converted.

Of course, for expository purposes, we have used an artificially simplified example. We also deliberately seeded some errors for LP to find. However, most of the errors discussed above occurred unintentionally as we wrote the example, and we did not notice them until we actually attempted the mechanical proofs.

7.8   Perspective

The Larch Shared Language includes several facilities for introducing checkable redundancy into specifications. These facilities were chosen to expose common classes of errors. They give specifiers a better chance of receiving diagnostics about specifications with unintended meanings, in much the same way that type systems give programmers a better chance of receiving diagnostics about erroneous programs.

A primary goal of Larch is to provide useful feedback to specifiers when there is something wrong with a specification. Hence we designed LP primarily as a debugging tool. We are not overly troubled that detecting inconsistencies is generally quicker and easier than certifying consistency.

We expect to discover flaws in specifications by having attempted proofs fail. LP does not automatically apply backwards inference techniques, and it requires more user guidance than some other provers. Much of this guidance is highly predictable, e.g, proving the hypotheses of deduction rules as lemmas. Although it is tempting to supply LP with heuristics that would generate such lemmas automatically, we feel that it is better to leave the guidance to the user. At many points in a proof, many different heuristics could apply. In our experience, choosing the next step in a proof (e.g., a case split or proof by induction)—or deciding that the proof attempt should be abandoned—often depends upon knowledge of the application. LP cannot reasonably be expected to possess this knowledge, especially when we are searching for a counterexample to a conjecture, rather than attempting to prove it. However, in some cases, the LSL Checker may be able to use the structure of LSL specifications to generate some of the guidance (e.g., using induction to prove a converts clause) that users must currently provide to LP.

The checkable redundancy that LSL encourages in specifications also supports regression testing as specifications evolve. When we change part of a specification (e.g., to strengthen or weaken the assertions of one trait), it is important to ensure that the change does not have unintended side-effects. LP's facilities for detecting inconsistencies help us discover grossly erroneous changes. Claims about other traits in the specification, which imply or assume the changed trait, can help us discover more subtle problems. If some of these claims have already been checked, LP's facilities for replaying proof scripts make it easy to recheck their proofs after a change—an important activity, but one that is likely to be neglected without mechanical assistance.

Chapter 8

Conclusion

Larch is still very much a "work in progress." New Larch interface languages are being designed, new tools are being built, and the existing languages and tools are in a state of evolution. Most significantly, specifications are being written.

But Larch has reached a divide, what Churchill might have called "the end of the beginning." Until now, most of the work on Larch has been done by the authors of this book and their close associates. We hope that the First International Workshop on Larch [66] and the publication of this book mark the beginning of the period when most Larch research, development, and application will be done by people we do not yet know.

THE ESSENCE OF LARCH

Over the years, we have spent many pages describing Larch languages, tools, and applications. However, the essence of Larch rests in a few principles that have guided our efforts:

- The most important use for specification is as a tool for helping to understand and document interfaces. Therefore, clarity is more important than any other property.

- Specifications should not just describe mathematical abstractions, but real interfaces supplied by programs. They should be written at the level of abstraction at which clients program. This usually means sinking to the level of a programming language.

- Structuring specifications into two tiers, which we have called the interface tier and the LSL tier, makes specifications easier to understand and facilitates reuse of parts of specifications.

  - The interface tier describes the observable behavior of program components. Since what a client can observe is likely to depend in fundamental ways on the client programming language, much can be gained by designing interface specification languages that are optimized for specific programming languages.

Specifications in this tier can be rather simple, provided that the right abstractions are provided in the LSL tier.

- The LSL tier describes mathematical abstractions that are independent of the details of any programming model. These are the principal reusable components of specifications. While we have used only one language (LSL) to write specifications in this tier, there is no fundamental reasons other languages could not be used. Languages used in this tier should have a simple semantics; they need not deal with messy issues such as runtime errors, which are better handled in the interface tier.

• Specification languages should be carefully designed. Having an elegant semantics is not enough. Careful attention to syntax and static semantic checking is crucial.

• Tool support is vital. One of the great virtues of using a formal notation is that tools can be used to help detect and isolate a variety of errors. Whenever we have improved our tools to detect a new class of errors, we have found more errors in existing specifications.

• Tools for checking interface specifications should be integrated with other programming language tools, e.g., preprocessors that enforce programming conventions.

• Specification must not be viewed as an isolated activity. It must be integrated with good programming practice. The goal is not to specify arbitrary programs, but to use specifications to help design, implement, document, and maintain good programs. Specifications can help in structuring these activities.

A CAUTIONARY NOTE

Throughout this book we have stressed ways in which formal specification can be used to help in building high quality software. However, we have tried not lose sight of the fact that formal specification is not a panacea. Good engineering practice is essential. To quote an anonymous referee of an early draft of this book,

> ...bullishness about formal methods must be strongly tempered by the following important realization: *Formalization should be aimed at achieving conceptual clarity, rather than*

as a mere exercise in encoding pieces of mathematics. No notation or toolset, however fancy and elaborate, can be a substitute for clear thought. At best, formalization can help clarify ideas and concepts by making them more tangible. At worst, poor or faulty formalization can cloud and confuse issues beyond repair.

Appendix A

An LSL Handbook

A.1 Introduction

This handbook supersedes Piece IV of *Larch in Five Easy Pieces* [51] and "A Larch Shared Language Handbook" [46].

READING THE HANDBOOK

This handbook contains a collection of traits written in LSL 2.4 that can be studied to learn more about LSL. Many traits are also suitable for use as specification components. They constitute a library for the LCL and LM3 tools; we hope that they will save others from reinventing wheels—especially polygonal ones. Other traits are more likely to be used as models for the development of similar specialized specification components.

This handbook is representative rather than complete. The LSL tier is open-ended because we believe that no handbook or library will ever include everything that will be needed. Users are encouraged to augment this handbook with additional traits, and to prepare their handbooks for particular applications.

This is not a textbook on discrete mathematics. If you already understand a collection of concepts (e.g., integer arithmetic), their formalization should make sense to you. If you don't, you should still be able to understand precisely what the definitions say (or don't say), but you probably won't get many clues as to why the particular definitions in (say) `Lattice` or `AbelianMonoid` are interesting and useful. Think of this handbook as the "collected formulas" that might appear as an appendix to a mathematics text.

There are many trade-offs in developing this kind of handbook:

- simplicity versus completeness,

- structure (include trait by reference) versus explicitness (copy trait),

- brevity versus explicit indication of consequences,

- concise versus mnemonic names,

- stylistic consistency versus an illustrative range of valid styles,

- standardization (for communication) versus flexibility (for efficiency in particular cases),

- selection among competing notations and definitions for concepts,

- conceptual elegance versus practical utility.

We expect that, in the not-too-distant future, specification handbooks will most often be used in their online forms, with browsing tools that enable readers to make many of these choices dynamically, according to their needs and preferences. Unfortunately, this book is still a hostage to the tyranny of paper, so we've had to make these choices in advance. There are general tendencies in the choices exhibited here, but we haven't applied any of our own guidelines slavishly. Many of the stylistic variations are intentional, but there are probably others that we simply didn't notice.

This handbook does not have to be read front-to-back. There is no "correct" order in which to study the traits. Feel free to browse and skip according to your interests and needs. Early sections tend to deal with specific constructs that occur frequently in program interface specifications, while later sections are somewhat more abstract, providing mathematical building blocks that can be used to define, classify, or generalize such constructs. When there didn't seem to be any natural order for things, we fell back on alphabetical order.

Traits in sections labeled *data types* or *data structures* are quite likely to be used directly in interface specifications. Traits in sections labeled *assumptions and implications* or *operator definitions* are more likely to be used in other traits.

Traits are listed in the index. If you don't know exactly what a referenced trait contains, you can always look it up. However, we have tried to use familiar names for familiar concepts. Particularly on first reading, it is probably better to assume that traits such as `Integer` and `TotalOrder` mean what you expect, than to flip continually from trait to trait and section to section.

An `implies` clause does not contribute to the meaning (i.e., the theory) of a legal trait. It is perfectly acceptable to ignore them, and it is often best to do so on first reading. However, they do offer you a chance to check your understanding, by giving examples of facts that are consequences of the definitions in the trait. They may also include alternative (and perhaps more familiar) definitions, or show connections that may not be obvious from looking at just the definitions in the traits.

Both `includes` and `assumes` clauses add axioms from referenced traits. They both have the same semantics within a trait in which they appear, so it's fine to ignore the distinction on first reading. But `assumes` clauses impose an additional proof obligation whenever the trait containing them is referenced in another trait, so they become very relevant when using traits to compose specifications.

Many abstract types are defined in two traits, one of which defines only the essential operators that characterize the type, while the other includes definitions for a richer set of operators in terms of the essential operators. The former kind of trait tends to be used in `assumes` and `implies` clauses; the latter, in `includes` clauses and in interface specifications. Compare, for example, `SetBasics` and `Set`, or `RelationBasics` and `Relation`.

Many traits include `Integer` and use sort `Int` where it might seem that `Natural` and `Nat` would be more natural choices—and, in some cases, would lead to somewhat simpler specifications. This is a consequence of the decision in the interface languages to base all the whole-number types on `Int`. The trait `IntegerPredicates` defines predicates to test for several commonly-used subsets of the integers. The alternative was a large amount of sort-conversion that would severely distract from the clarity of interface specifications. So we pay a small price in the LSL tier for greater simplicity in the interface tier.

If a definition seems "unnatural" to you, you will find it instructive to try to construct a more natural definition yourself. If you find one, you will have gained some experience in writing LSL specifications; if you don't, you may have gained some insight into the reason for the "unnatural" definition.

The traits in this handbook have passed the scrutiny of the LSL Checker, which parses, expands trait references, resolves overloading, and sort-checks. Most of them have not yet been subjected to additional checking of the kind described in Chapter 7.

The online version of this handbook is still evolving. The authors would appreciate all kinds of feedback from readers and users. Are there errors or sources of confusion? Have we omitted something that would be widely useful? Are there better ways to define some of the concepts?

NAMING AND LEXICAL CONVENTIONS

Sort names:

- Numeric types: Int for integers, P for positive numbers, Q for rationals, F for floating point, and N otherwise.

- T if there is only one "interesting" sort in the trait.

- Container traits: E for elements, C for containers.

Operator names:

- o for a generic infix operator and also for the composition of maps and relations.

- ◇ for a generic relation.

For convenience in manipulating the online form of the handbook, we have chosen a sequence of ISO Latin characters to represent each non-ISO Latin symbol used in the handbook. Some of them are chosen for visual similarity (e.g., → is written as -> and ≤ is written as <=); others have been modeled on TeX's choices (e.g., o is written as \circ and ∈ is written as \in). A complete list is given in Section C.

Each Larch interface language defines its own notation for literals, based on the programming language's notation; numerical types will generally include the trait schema DecimalLiterals.

Many traits have a size or count operator whose value is always non-negative. For reasons given in the previous section, except within Section A.15, Number theory, we have given their range as Int, from trait Integer, rather than as N, from trait Natural.

A.2   Foundations

DATA TYPE: BOOLEAN

```
Boolean: trait
  % This trait is given for documentation only.
  % It is implicit in LSL.
  introduces
    true, false: → Bool
    ¬__: Bool → Bool
    __∧__, __∨__, __⇒__: Bool, Bool → Bool
  asserts
    Bool generated by true, false
    ∀ b: Bool
      ¬ true == false;
      ¬ false == true;
      true ∧ b == b;
      false ∧ b == false;
      true ∨ b == true;
      false ∨ b == b;
      true ⇒ b == b;
      false ⇒ b == true
  implies
    AC (∧ , Bool),
    AC (∨, Bool),
    Distributive (∨ for +, ∧ for *, Bool for T),
    Distributive (∧ for +, ∨ for *, Bool for T),
    Involutive (¬__, Bool),
    Transitive (⇒ for ◇, Bool for T)
    ∀ b1, b2, b3: Bool
      ¬(b1 ∧ b2) == ¬b1 ∨ ¬b2;
      ¬(b1 ∨ b2) == ¬b1 ∧ ¬b2;
      b1 ∨ (b1 ∧ b2) == b1;
      b1 ∧ (b1 ∨ b2) == b1;
      b2 ∨ ¬b2;
      (b1 = b2) ∨ (b1 = b3) ∨ (b2 = b3);
      b1 ⇒ b2 == ¬b1 ∨ b2
```

OPERATOR DEFINITION: IF THEN ELSE

```
Conditional (T): trait
  % This trait is given for documentation only.
  % It is implicit in LSL.
  introduces if__then__else__: Bool, T, T → T
  asserts
    ∀ x, y, z: T
      if true then x else y == x;
      if false then x else y == y
  implies ∀ b: Bool, x: T
    if b then x else x == x
```

A.3   Integers

DATA TYPE

```
Integer (Int): trait
  % The usual (unbounded) integers operators
  includes
    DecimalLiterals (Int for N),
    TotalOrder (Int)
  introduces
    0, 1: → Int
    succ, pred, -__, abs: Int → Int
    __+__, __-__, __*__: Int, Int → Int
    div, mod, min, max: Int, Int → Int
  asserts
    Int generated by 0, succ, pred
    ∀ x, y: Int
      succ(pred(x)) == x;
      pred(succ(x)) == x;
      -0 == 0;
      -succ(x) == pred(-x);
      -pred(x) == succ(-x);
      abs(x) == max(-x, x);
      x + 0 == x;
      x + succ(y) == succ(x + y);
      x + pred(y) == pred(x + y);
      x - y == x + (-y);
      x * 0 == 0;
      x*succ(y) == (x*y) + x;
      x*pred(y) == (x*y) - x;
      y > 0 ⇒ mod(x, y) + (div(x, y) * y) = x;
      y > 0 ⇒ mod(x, y) ≥ 0;
      y > 0 ⇒ mod(x, y) < y;
      min(x, y) == if x ≤ y then x else y;
      max(x, y) == if x ≤ y then y else x;
      x < succ(x)
  implies
    AC (+, Int),
    AC (*, Int),
    AC (min, Int),
    AC (max, Int),
    RingWithUnit (Int for T)
    Int generated by 1, +, -__:Int→Int
```

```
∀ x, y: Int
  x < y == succ(x) < succ(y);
  x ≤ y == x ≤ succ(y)
converts
  1, -__:Int→Int, __-__:Int,Int→Int,
  abs, +, *, div, mod, min, max, ≤, ≥, <, >
```

LITERALS

```
DecimalLiterals (N): trait
  % A built-in trait schema given here
  % for documentation only
  introduces
    0, 1, 2, 3, 4, 5, 6, 7, 8, 9, 10, 11 %, ...
    : → N
    succ: N → N
  asserts equations
    1 == succ(0);
    2 == succ(1);
    3 == succ(2);
  % ... as far as needed for any literals
  % of sort N appearing in the including trait
```

OPERATOR DEFINITIONS

```
IntegerPredicates (Int): trait
  % Frequently used subranges of the integers
  assumes Integer
  introduces
    InRange: Int, Int, Int → Bool
    Natural, Positive, Signed, Unsigned: Int → Bool
    maxSigned, maxUnsigned: → Int
  asserts forall n, low, high: Int
    InRange(n, low, high) == low ≤ n ∧ n ≤ high;
    Natural(n) == n ≥ 0;
    Positive(n) == n > 0;
    Signed(n) ==
      InRange(n, -succ(maxSigned), maxSigned);
    Unsigned(n) == InRange(n, 0, maxUnsigned)
  implies ∀ n: Int
    Positive(n) ⇒ Natural(n);
    Unsigned(n) ⇒ Natural(n)
```

A.4 Enumerations

```
Enumeration (T): trait
  % Enumeration, comparison, and ordinal position
  % operators, often used with "enumeration of"
  assumes Integer
  includes DerivedOrders
  introduces
    first, last: → T
    succ, pred: T → T
    ord: T → Int
    val: Int → T
  asserts
    T generated by first, succ
    T generated by last, pred
    ∀ x, y: T
      ord(first) == 0;
      x ≠ last ⇒ ord(succ(x)) = ord(x) + 1;
      x ≠ last ⇒ pred(succ(x)) = x;
      val(ord(x)) == x;
      x ≤ y == ord(x) ≤ ord(y);
      x ≤ last
  implies
    TotalOrder
    T generated by val
    T partitioned by ord
    ∀ x: T
      x ≠ first ⇒ succ(pred(x)) = x;
      x ≠ last ⇒ x < succ(x);
      first ≤ x;
      ord(x) ≥ 0
    converts
      first:→T, succ:T→T, pred:T→T, ord,
        ≤:T,T→Bool, ≥:T,T→Bool,
        <:T,T→Bool, >:T,T→Bool
      exempting succ(last), pred(first)
```

A.5   Containers

Throughout this section we use E for the element sort, and C for the container sort. This simplifies comparisons among data structures and makes it easier to write generic operator definitions that work for several kinds of containers. Since variable names are local to traits, we imposed no such uniformity on them.

UNORDERED DATA STRUCTURES

```
SetBasics (E, C): trait
  % Essential finite-set operators
  introduces
    { }: → C
    insert: E, C → C
    __ ∈ __: E, C → Bool
  asserts
    C generated by { }, insert
    C partitioned by ∈
    ∀ s: C, e, e1, e2: E
      ¬ (e ∈ { });
      e1 ∈ insert (e2, s) == e1 = e2 ∨ e1 ∈ s
  implies
    InsertGenerated ({ } for empty)
    ∀ e, e1, e2: E, s: C
      insert (e, s) ≠ { };
      insert (e, insert (e, s)) == insert (e, s);
      insert (e1, insert (e2, s)) ==
        insert (e2, insert (e1, s))
    converts ∈
```

```
Set (E, C): trait
  % Common set operators
  includes
    SetBasics,
    Integer,
    DerivedOrders (C, ⊆ for ≤, ⊇ for ≥,
                      ⊂ for <, ⊃ for >)
  introduces
    __∉ __: E, C → Bool
    delete: E, C → C
    {__}: E → C
    __ ∪ __, __ ∩ __, __-__: C, C → C
    size: C → Int
  asserts
    ∀ e, e1, e2: E, s, s1, s2: C
      e ∉ s == ¬(e ∈ s);
      { e } == insert(e, {});
      e1 ∈ delete(e2, s) == e1 ≠ e2 ∧ e1 ∈ s;
      e ∈ (s1 ∪ s2) == e ∈ s1 ∨ e ∈ s2;
      e ∈ (s1 ∩ s2) == e ∈ s1 ∧ e ∈ s2;
      e ∈ (s1 - s2)  == e ∈ s1 ∧ e ∉ s2;
      size({}) == 0;
      size(insert(e, s)) ==
        if e ∉ s then size(s) + 1 else size(s);
      s1 ⊆ s2 == s1 - s2 = {}
  implies
    AbelianMonoid (∪ for ∘, {} for unit, C for T),
    AC (∩, C),
    JoinOp (∪, {} for empty),
    MemberOp ({} for empty),
    PartialOrder (C, ⊆ for ≤, ⊇ for ≥,
                      ⊂ for <, ⊃ for >)
    C generated by {}, {__}, ∪
    ∀ e: E, s, s1, s2: C
      s1 ⊆ s2 ⇒ (e ∈ s1 ⇒ e ∈ s2);
      size(s) ≥ 0
    converts
      ∈, ∉, {__}, delete, size, ∪, ∩, -:C,C→C,
      ⊆, ⊇, ⊂, ⊃
```

```
BagBasics (E, C): trait
  % Essential bag operators
  includes Integer
  introduces
    { }: → C
    insert: E, C → C
    count: E, C → Int
  asserts
    C generated by { }, insert
    C partitioned by count
    ∀ b: C, e, e1, e2: E
      count(e, { }) == 0;
      count(e1, insert(e2, b)) ==
        count(e1, b) + (if e1 = e2 then 1 else 0)
  implies
    InsertGenerated ({ } for empty)
    ∀ e: E, b: C
      insert(e, b) ≠ { };
      count(e, b) ≥ 0
    converts count
```

```
Bag (E, C): trait
  % Common bag operators
  includes
    BagBasics,
    DerivedOrders (C, ⊆ for ≤, ⊇ for ≥,
                   ⊂ for <, ⊃ for >)
  introduces
    delete: E, C → C
    {__}: E → C
    __∈__, __∉__: E, C → Bool
    size: C → Int
    __∪__, __-__: C, C → C
  asserts
    ∀ e, e1, e2: E, b, b1, b2: C
      count(e1, delete(e2, b)) ==
        if e1 = e2 then max(0, count(e1, b) - 1)
        else count(e1, b);
      { e } == insert(e, { });
      e ∈ b == count(e, b) > 0;
      e ∉ b == count(e, b) = 0;
      size({ }) == 0;
      size(insert(e, b)) == size(b) + 1;
      count(e, b1 ∪ b2) ==
        count(e, b1) + count(e, b2);
      count(e, b1 - b2) ==
        max(0, count(e, b1) - count(e, b2));
      b1 ⊆ b2 == b1 - b2 = { };
  implies
    AbelianMonoid (∪ for ∘, { } for unit, C for T),
    JoinOp (∪, { } for empty),
    MemberOp ({ } for empty),
    PartialOrder (C, ⊆ for ≤, ⊇ for ≥,
                  ⊂ for <, ⊃ for >)
    ∀ e, e1, e2: E, b, b1, b2: C
      insert(e, b) ≠ { };
      count(e, b) ≥ 0;
      count(e, b) ≤ size(b);
      b1 ⊆ b2 ⇒ count(e, b1) ≤ count(e, b2)
    converts count, ∈, ∉, {__}, ∪, -:C,C→C,
      delete, size, ⊆, ⊇, ⊂, ⊃
```

INSERTION ORDERED DATA STRUCTURES

```
StackBasics (E, C): trait
  % Essential LIFO operators
  includes Integer
  introduces
    empty: → C
    push: E, C → C
    top: C → E
    pop: C → C
  asserts
    C generated by empty, push
    ∀ e: E, stk: C
      top(push(e, stk)) == e;
      pop(push(e, stk)) == stk;
  implies converts top, pop
    exempting top(empty), pop(empty)

Stack (E, C): trait
  % Common LIFO operators
  includes StackBasics, Integer
  introduces
    count: E, C → Int
    __ ∈ __: E, C → Bool
    size: C → Int
    isEmpty: C → Bool
  asserts
    ∀ e: E, stk: C
      size(empty) == 0;
      size(push(e, stk)) == size(stk) + 1;
      isEmpty(stk) == stk = empty
  implies
    Container (push for insert, top for head,
               pop for tail)
    C partitioned by top, pop, isEmpty
    ∀ stk: C
      size(stk) ≥ 0
    converts top, pop, count, ∈, size, isEmpty
      exempting top(empty), pop(empty)
```

```
Queue (E, C): trait
  % FIFO operators
  includes Integer
  introduces
    empty: → C
    append: E, C → C
    count: E, C → Int
    __ ∈ __: E, C → Bool
    head: C → E
    tail: C → C
    len: C → Int
    isEmpty: C → Bool
  asserts
    C generated by empty, append
    ∀ q: C, e, e1: E
      count(e, empty) == 0;
      count(e, append(e1, q)) ==
        count(e, q) + (if e = e1 then 1 else 0);
      e ∈ q == count(e, q) > 0;
      head(append(e, q)) ==
        if q = empty then e else head(q);
      tail(append(e, q)) ==
        if q = empty then empty
        else append(e, tail(q));
      len(empty) == 0;
      len(append(e, q)) == len(q) + 1;
      isEmpty(q) == q = empty
  implies
    Container (append for insert)
    C partitioned by head, tail, isEmpty
    ∀ q: C
      len(q) ≥ 0
    converts head, tail, len
      exempting head(empty), tail(empty)
```

```
Deque (E, C): trait
  % Double ended queue operators
  includes Integer
  introduces
    empty: → C
    __ ⊣ __ : E, C → C
    __ ⊢ __ : C, E → C
    count: E, C → Int
    __ ∈ __: E, C → Bool
    head, last: C → E
    tail, init: C → C
    len: C → Int
    isEmpty: C → Bool
  asserts
    C generated by empty, ⊢
    ∀ e, e1, e2: E, d: C
      count(e, empty) == 0;
      count(e, e1 ⊣ d) ==
        count(e, d) + (if e = e1 then 1 else 0);
      e ∈ d == count(e, d) > 0;
      e ⊣ empty == empty ⊢ e;
      (e1 ⊣ d) ⊢ e2 == e1 ⊣ (d ⊢ e2);
      head(e ⊣ d) == e;
      last(d ⊢ e) == e;
      tail(e ⊣ d) == d;
      init(d ⊢ e) == d;
      len(empty) == 0;
      len(d ⊢ e) == len(d) + 1;
      isEmpty(d) == d = empty
  implies
    Stack (head for top, tail for pop,
           ⊣ for push, len for size),
    Queue (⊣ for append, last for head,
           init for tail)
    C generated by empty, ⊣
    C partitioned by len, head, tail
    C partitioned by len, last, init
    ∀ d: C
      d ≠ empty
        ⇒ (head(d) ⊣ tail(d) = d
             ∧ init(d) ⊢ last(d) = d)
    converts head, last, tail, init, len
      exempting head(empty), last(empty),
        tail(empty), init(empty)
```

```
List (E, C): trait
  % Add singleton and concatenation
  includes Deque
  introduces
    {__}: E → C
    __ || __ : C, C → C
  asserts ∀ e: E, ls, ls1, ls2: C
    {e} == empty ⊢ e;
    ls || empty == ls;
    ls1 || (ls2 ⊢ e) == (ls1 || ls2) ⊢ e
  implies
    C generated by empty, {__}, ||
    converts head, last, tail, init, len, {__}, ||
      exempting head(empty), last(empty),
        tail(empty), init(empty)

String (E, C): trait
  % Index, substring
  includes List
  introduces
    __[__]: C, Int → E
    prefix: C, Int → C
    removePrefix: C, Int → C
    substring: C, Int, Int → C
  asserts ∀ e: E, s: C, i, n: Int
    tail(empty) == empty;
    init(empty) == empty;
    s[0] == head(s);
    n ≥ 0 ⇒ s[n + 1] = tail(s)[n];
    prefix(empty, n) == empty;
    prefix(s, 0) == empty;
    n ≥ 0
      ⇒ prefix(e ⊣ s, n + 1) = e ⊣ prefix(s, n);
    removePrefix(s, 0) == s;
    n ≥ 0
      ⇒ removePrefix(s, n + 1)
        = removePrefix(tail(s), n);
    substring(s, 0, n) == prefix(s, n);
    i ≥ 0
      ⇒ substring(s, i + 1, n)
        = substring(tail(s), i, n)
  implies
    IndexOp (⊣ for insert)
    C partitioned by len, __[__]
    converts tail, init
```

```
Sequence (E, C): trait
  % Comparison, subsequences
  assumes StrictPartialOrder (>, E)
  includes
    LexicographicOrder,
    String
  introduces
    isPrefix, isSubstring, isSubsequence: C, C → Bool
    find: C, C → Int
  asserts ∀  e, e1, e2: E, s, s1, s2: C
    isPrefix(s1, s2) == s1 = prefix(s2, len(s1));
    isSubstring(s1, s2) ==
      isPrefix(s1, s2) ∨ isSubstring(s1, tail(s2));
    isSubsequence(empty, s);
    ¬isSubsequence(e ⊣ s, empty);
    isSubsequence(e1 ⊣ s1, e2 ⊣ s2) ==
      (e1 = e2 ∧ isSubsequence(s1, s2))
        ∨ isSubsequence(e1 ⊣ s1, s2);
    find(s1, s2) ==
      if isPrefix(s1, s2) then 0
      else find(s1, tail(s2)) + 1
  implies
    IsPO (isPrefix, C),
    IsPO (isSubstring, C),
    IsPO (isSubsequence, C)
  ∀ s, s1, s2: C, i, n: Int
    isPrefix(prefix(s, n), s);
    isSubstring(substring(s, i, n), s);
    isSubstring(s1, s2) ⇒ isSubsequence(s1, s2)
  converts
      isPrefix, isSubstring, isSubsequence, find
    exempting ∀ s: C, e: E   find(e ⊣ s, empty)
```

CONTENT ORDERED DATA STRUCTURES

```
PriorityQueue (>:E,E→Bool, E, C): trait
  % Enumerate by order on elements
  assumes TotalOrder (E for T)
  includes Integer
  introduces
    empty: → C
    add: E, C → C
    count: E, C → Int
    __ ∈ __: E, C → Bool
    head: C → E
    tail: C → C
    len: C → Int
    isEmpty: C → Bool
  asserts
    C generated by empty, add
    C partitioned by head, tail, isEmpty
    ∀ e, e1: E, q: C
      count(e, empty) == 0;
      count(e, add(e1, q)) ==
        count(e, q) + (if e = e1 then 1 else 0);
      e ∈ q == count(e, q) > 0;
      head(add(e, q)) ==
        if q = empty ∨ e > head(q) then e
        else head(q);
      tail(add(e, q)) ==
        if q = empty ∨ e > head(q) then q
        else add(e, tail(q));
      len(empty) == 0;
      len(add(e, q)) == len(q) + 1;
      isEmpty(q) == q = empty
  implies
    Container (add for insert)
    ∀ e, e1, e2: E, q: C
      add(e1, add(e2, q)) = add(e2, add(e1, q));
      len(q) ≥ 0;
      add(e, q) ≠ empty
    converts count, ∈, head, tail, len, isEmpty
      exempting head(empty), tail(empty)
```

```
ChoiceSet (E, C): trait
   % A set with a weakly-specified choose operator
   includes Set
   introduces
      choose: C → E
      rest: C → C
      isEmpty: C → Bool
   asserts ∀ e, el: E, s: C
      s ≠ {} ⇒ choose(s) ∈ s;
      s ≠ {} ⇒ rest(s) = delete(choose(s), s);
      isEmpty(s) == s = {}
   implies
      C partitioned by choose, rest, isEmpty
      ∀ e: E, s: C
         s ≠ {} ⇒ s = insert(choose(s), rest(s))

ChoiceBag (E, C): trait
   % A bag with a weakly-specified choose operator
   includes Bag
   introduces
      choose: C → E
      rest: C → C
      isEmpty: C → Bool
   asserts ∀ e, el: E, b: C
      b ≠ {} ⇒ choose(b) ∈ b;
      b ≠ {} ⇒ rest(b) = delete(choose(b), b);
      isEmpty(b) == b = {}
   implies
      Container (choose for head, rest for tail,
                 {} for empty)
      C partitioned by choose, rest, isEmpty
      ∀ e: E, b: C
         b ≠ {} ⇒ b = insert(choose(b), rest(b))
```

ASSUMPTIONS AND IMPLICATIONS

```
InsertGenerated (E, C): trait
  % C's contain finitely many E's
  introduces
    empty: → C
    insert: E, C → C
  asserts
    C generated by empty, insert

Container (E, C): trait
  % head and tail enumerate contents of a C
  includes InsertGenerated, Integer
  introduces
    isEmpty: C → Bool
    count: E, C → Int
    __ ∈ __: E, C → Bool
    head: C → E
    tail: C → C
  asserts
    C partitioned by isEmpty, head, tail
    ∀ e, e1: E, c: C
      isEmpty(empty);
      ¬isEmpty(insert(e, c));
      count(e, empty) == 0;
      count(e, insert(e1, c)) ==
        count(e, c) + (if e = e1 then 1 else 0);
      e ∈ c == count(e, c) > 0;
      ¬isEmpty(c) ⇒
        count(e, insert(head(c), tail(c)))
          = count(e, c)
  implies
    ∀ c: C
      ¬isEmpty(c) ⇒ count(head(c), c) > 0;
    converts isEmpty, count, ∈
```

OPERATOR DEFINITIONS

```
MemberOp: trait
  assumes InsertGenerated
  introduces
    __ ∈ __, __ ∉ __: E, C → Bool
  asserts ∀ e, e1, e2: E, c: C
    e ∉ c == ¬ (e ∈ c);
    e ∉ empty;
    e1 ∈ insert(e2, c) == e1 = e2 ∨ e1 ∈ c
  implies converts ∈, ∉

JoinOp (⋈): trait
  % Container combining operator
  % e.g., union, concatenation
  assumes InsertGenerated
  introduces __ ⋈ __: C, C → C
  asserts ∀ e: E, c, c1, c2: C
    empty ⋈ c == c;
    insert(e, c1) ⋈ c2 == insert(e, c1 ⋈ c2)
  implies
    Associative (⋈, C)
    converts ⋈

ReverseOp: trait
  % An operator on lists commonly used
  % to demonstrate theorem provers.
  assumes List
  introduces reverse: C → C
  asserts ∀ e: E, l, l1, l2: C
    reverse(empty) == empty;
    reverse(e ⊣ l) == reverse(l) ⊢ e
  implies ∀ e: E, l, l1, l2: C
      reverse(reverse(l)) == l;
      l ≠ empty ⇒ head(reverse(l)) = last(l);
      l ≠ empty
        ⇒ tail(reverse(l)) = reverse(init(l));
      len(reverse(l)) == len(l);
      reverse(l1 ∥ l2) == reverse(l2) ∥ reverse(l1)
    converts reverse
```

```
IndexOp: trait
  % Select the i-th  element in the container
  % (in enumeration order).
  assumes Integer, Container
  introduces __[__]: C, Int → E
  asserts ∀ c: C, i: Int
    c[0] == head(c);
    i ≥ 0 ⇒ c[i+1] = tail(c)[i]
```

CoerceContainer (DC, RC) defines an operator to convert from
a term of one container sort, DC, to a term of another container sort, RC,
with the same elements inserted in the same order. For example, a stack can
be mapped to a queue. More interestingly, a list can be mapped to a bag, or
a bag to a set; these mappings lose information on order and on multiplicity,
respectively, so the inverse mappings would introduce inconsistencies.

```
CoerceContainer (DC, RC): trait
  % Insert each element of DC in a new RC
  assumes
    InsertGenerated (DC for C),
    InsertGenerated (RC for C)
  introduces coerce: DC → RC
  asserts ∀ dc: DC, e: E
    coerce(empty) == empty;
    coerce(insert(e, dc)) == insert(e, coerce(dc))
  implies converts coerce

Permutation (E, C): trait
  % Test for having the same elements
  assumes Container
  includes
    Bag (B for C),
    CoerceContainer (C for DC, B for RC)
  introduces isPermutation: C, C → Bool
  asserts ∀ c1, c2: C
      isPermutation(c1, c2) == coerce(c1) = coerce(c2)
  implies ∀ e: E, c1, c2: C
    isPermutation(c1, c2)
      ⇒ count(e, c1) = count(e, c2)
```

The following traits "promote" various operators on elements to corresponding operators on containers.

```
ElementTest (pass, E, C, T); trait
  % filter collects elements accepted by pass
  assumes InsertGenerated
  introduces
    pass: E, T → Bool
    filter: C, T → C
    allPass: C, T → Bool
    somePass: C, T → Bool
  asserts ∀ c: C, e: E, t: T
    filter(empty, t) == empty;
    filter(insert(e, c), t) ==
        if pass(e, t) then insert(e, filter(c, t))
        else filter(c, t);
    allPass(empty, t);
    allPass(insert(e, c), t) ==
        pass(e, t) ∧ allPass(c, t);
    somePass(c, t) == filter(c, t) ≠ empty
  implies converts filter, somePass, allPass

PairwiseExtension (o, ⊙, E, C): trait
  % Induce a binary operator on containers
  % from a binary operator on elements.
  assumes Container (E, C)
  introduces
    __ o __ : E, E → E
    __ ⊙ __ : C, C → C
  asserts ∀ e1, e2: E, c1, c2: C
    empty ⊙ empty == empty;
    (c1 ≠ empty ∧ c2 ≠ empty)
        ⇒ c1 ⊙ c2 = insert(head(c1) o head(c2),
                            tail(c1) ⊙ tail(c2));
  implies converts ⊙
    exempting ∀ e: E, c: C
        empty ⊙ insert(e, c), insert(e, c) ⊙ empty
```

```
PointwiseImage: trait
  % Apply elemOp to each element
  assumes
    InsertGenerated (DE for E, DC for C),
    InsertGenerated (RE for E, RC for C)
  introduces
    elemOp: DE → RE
    containerOp: DC → RC
  asserts ∀ dc: DC, de: DE
    containerOp(empty) == empty;
    containerOp(insert(de, dc)) ==
      insert(elemOp(de), containerOp(dc))
  implies converts containerOp

ReduceContainer (unit, o): trait
  % Insert the operator in enumeration order.
  assumes Container
  introduces
    unit: → E
    __ o __: E, E → E
    reduce: C → E
  asserts ∀ c: C
    reduce(c) ==
      if c = empty then unit
      else head(c) o reduce(tail(c))
  implies converts reduce
```

A.6   Branching structures

DATA STRUCTURES

The following trait defines the operators on a list (of sort C), each of whose elements (of sort E) is either an atom (of sort A) or a list.

```
ListStructure (A, E, C): trait
  % Classical LISP
  includes List
  E union of list: C, atom: A

BinaryTree (E, T): trait
  % One of the many interesting tree structures
  introduces
    [__]: E → T
    [__, __]: T, T → T
    content: T → E
    first, second: T → T
    isLeaf: T → Bool
  asserts
    T generated by [__], [__, __]
    T partitioned by content, first, second, isLeaf
    ∀ e: E, t1, t2: T
      content([e]) == e;
      first([t1, t2]) == t1;
      second([t1, t2]) == t2;
      isLeaf([e]);
      ¬isLeaf([t1, t2])
  implies converts isLeaf
```

OPERATOR DEFINITIONS

```
ListStructureOps (A, E, C): trait
  % Operators frequently used in
  % theorem proving demonstrations.
  assumes ListStructure
  introduces
    flatten, reverseAll: C → C
    countAtoms: C → Int
  asserts ∀ a: A, l, l1, l2: C
    flatten(empty) == empty;
    flatten(atom(a) ⊣ l) == atom(a) ⊣ flatten(l);
    flatten(list(l1) ⊣ l2) ==
      flatten(l1) ‖ flatten(l2);
    reverseAll(empty) == empty;
    reverseAll(atom(a) ⊣ l) ==
      reverseAll(l) ⊢ atom(a);
    reverseAll(list(l1) ⊣ l2) ==
      reverseAll(l2) ⊢ list(reverseAll(l1));
    countAtoms(l) == len(flatten(l))
  implies
    ∀ l, l1, l2: C
      flatten(l1 ‖ l2) == flatten(l1) ‖ flatten(l2);
      flatten(flatten(l)) == flatten(l);
      reverseAll(l1 ‖ l2) ==
        reverseAll(l2) ‖ reverseAll(l1);
      reverseAll(flatten(l)) ==
        flatten(reverseAll(l));
      reverseAll(reverseAll(l)) == l;
      countAtoms(l1 ‖ l2) ==
        countAtoms(l1) + countAtoms(l2);
      countAtoms(flatten(l)) == countAtoms(l);
      countAtoms(reverseAll(l)) == countAtoms(l)
    converts flatten, reverseAll, countAtoms
```

A.7   Maps

DATA STRUCTURES

Arrays are heavily-used data structures; programming languages often provide a large number of operators. The following definitions are only a sample.

```
Array1 (E, I, A): trait
  % Basic one-dimensional array operators
  introduces
    assign: A, I, E → A
    __[__]: A, I → E
  asserts
    ∀ a: A, i, j: I, e: E
      assign(a, i, e)[j] ==
        if i = j then e else a[j]

Array2 (E, I1, I2, A): trait
  % Basic two-dimensional array operators
  introduces
    assign: A, I1, I2, E → A
    __[__, __]: A, I1, I2 → E
  asserts
    ∀ a: A, i1, j1: I1, i2, j2: I2, e: E
      assign(a, i1, i2, e)[j1, j2] ==
        if i1 = j1 ∧ i2 = j2 then e else a[j1, j2]

ArraySlice2 (E, I1, I2, A): trait
  % A two-dimensional array
  % treated as a vector of vectors
  includes
    Array1 (E, I2, A1),
    Array1 (A1, I1, A)
  introduces
    assign: A, I1, I2, E → A
    __[__, __]: A, I1, I2 → E
  asserts
    ∀ a: A, i1: I1, i2: I2, e: E
      a[i1, i2] == (a[i1])[i2];
      assign(a, i1, i2, e) ==
        assign(a, i1, assign(a[i1], i2, e))
```

The maps of the following trait are finitely generated by {} and update.

```
FiniteMap (M, D, R): trait
  % An M is a map from D's to R's.
  introduces
    { }: → M
    update: M, D, R → M
    apply: M, D → R
    defined: M, D → Bool
  asserts
    M generated by { }, update
    M partitioned by apply, defined
    ∀ m: M, d, d1, d2: D, r: R
      apply(update(m, d2, r), d1) ==
        if d1 = d2 then r else apply(m, d1);
      ¬defined({ }, d);
      defined(update(m, d2, r), d1) ==
        d1 = d2 ∨ defined(m, d1)
  implies
    Array1 (update for assign, apply for __[__],
            M for A, D for I, R for E)
    converts apply, defined
      exempting ∀ d: D apply({ }, d)
```

OPERATOR DEFINITION

```
ComposeMaps (M1, M2, D, T, R): trait
  % If m1 is a map from D to T
  % and m2 is a map from T to R,
  % m1 o m2 is a map from D to R.
  assumes
    FiniteMap (M1, T, R),
    FiniteMap (M2, D, T)
  includes FiniteMap
  introduces __ o __: M1, M2 → M
  asserts ∀ m1: M1, m2: M2, d: D
    apply(m1 o m2, d) == apply(m1, apply(m2, d));
    defined(m1 o m2, d) ==
      defined(m2, d) ∧ defined(m1, apply(m2, d))
```

A.8   Relations

DATA STRUCTURE

The following traits do not presume that the domain sort, E, is generated
by any fixed set of operators. Subsets of E are represented by subrelations
of the identity relation.

```
Relation (E, R): trait
  includes
    RelationBasics,
    RelationOps,
    RelationPredicates

RelationBasics (E, R): trait
  % e1 ⟨ r ⟩ e2 means e1 is related to e2 by r.
  introduces
    __ ⟨ __ ⟩ __: E, R, E → Bool
    ⊥, ⊤, I: → R
    [__, __]: E, E → R
    -__, __⁻¹: R → R
    __ ∪ __: R, R → R
  asserts
    R partitioned by __ ⟨ __ ⟩ __
    ∀ e, e1, e2, e3, e4: E, r, r1, r2: R
      ¬(e1 ⟨ ⊥ ⟩ e2);
      e1 ⟨ ⊤ ⟩ e2;
      e1 ⟨ I ⟩ e2 == e1 = e2;
      e1 ⟨ [e2, e3] ⟩ e4 == e1 = e2 ∧ e3 = e4;
      e1 ⟨ -r ⟩ e2 == ¬(e1 ⟨ r ⟩ e2);
      e1 ⟨ r⁻¹ ⟩ e2 == e2 ⟨ r ⟩ e1;
      e1 ⟨ r1 ∪ r2 ⟩ e2 == e1 ⟨ r1 ⟩ e2 ∨ e1 ⟨ r2 ⟩ e2
  implies
    AbelianMonoid (⊥ for unit, ∪ for o, R for T),
    Involutive (__⁻¹, R),
    Involutive (-__, R)
    equations
      -⊥ == ⊤;
      -⊤ == ⊥;
      ⊥⁻¹ == ⊥;
      ⊤⁻¹ == ⊤
    converts ∪, -__, __⁻¹
```

OPERATOR DEFINITIONS

The `skolem` operator is introduced solely to get around the absence of existential quantifiers in LSL.

```
RelationOps: trait
  % Useful non-primitive operators on relations.
  assumes RelationBasics
  includes
    DerivedOrders (R, ⊆ for ≤, ⊇ for ≥,
                   ⊂ for <, ⊃ for >)
  introduces
    __ ∈ __, __ ∉ __: E, R → Bool
    set, dom, range, __⁺, __*: R → R
    __ ∩ __, __ ∘ __, __ - __, __ × __: R, R → R
    domRestrict, rangeRestrict, image: R, R → R
    skolem: E, R, R, E → E
  asserts
    ∀ e, e1, e2, e3: E, r, r1, r2: R
      e ∈ r == e ⟨ r ⟩ e;
      e ∉ r == ¬ (e ∈ r);
      set(r) == r ∩ I;
      dom(r) == set(r ∘ T);
      range(r) == set(T ∘ r);
      e1 ⟨ r1 ∩ r2 ⟩ e2 == e1 ⟨ r1 ⟩ e2 ∧ e1 ⟨ r2 ⟩ e2;
      (e1 ⟨ r1 ⟩ e2 ∧ e2 ⟨ r2 ⟩ e3)
        ⇒ e1 ⟨ r1 ∘ r2 ⟩ e3;
      e1 ⟨ r1 ∘ r2 ⟩ e2
        ⇒ (e1 ⟨ r1 ⟩ skolem(e1, r1, r2, e2)
          ∧ skolem(e1, r1, r2, e2) ⟨ r2 ⟩ e2);
      r⁺ == r ∘ (r*);
      r* == I ∪ (r⁺);
      (r1 = I ∪ r2 ∧ r2 = r ∘ r1) ⇒
        ((r*) ⊆ r1 ∧ (r⁺) ⊆ r2);
      r1 - r2 == r1 ∩ (-r2);
      r1 × r2 == set(r1) ∘ T ∘ set(r2);
      r1 ⊆ r2 == r1 - r2 = ⊥;
      domRestrict(r1, r2) == r1 ∩ (r2 ∘ T);
      image(r1, r2) == set(r1) ∘ r2;
      rangeRestrict(r1, r2) == r1 ∩ (T ∘ r2)
```

```
implies
   AbelianMonoid (⊤ for unit, ∩ for ∘, R for T),
   Distributive (∪ for +, ∩ for *, R for T),
   Distributive (∩ for +, ∪ for *, R for T),
   Idempotent (set, R),
   Monoid (I for unit, R for T),
   Lattice (R for T, ∪ for ⊔, ∩ for ⊓,
      ⊆ for ≤, ⊇ for ≥, ⊂ for <, ⊃ for >),
   PartialOrder (R, ⊆ for ≤, ⊇ for ≥, ⊂ for <,
      ⊃ for >)
   ∀ e: E, r, r1, r2: R
     e ∈ r == e ∈ set(r);
     -(r1 ∪ r2) == (-r1) ∩ (-r2);
     -(r1 ∩ r2) == (-r1) ∪ (-r2);
     (r1 ∘ r2)⁻¹ == (r2⁻¹) ∘ (r1⁻¹)
   converts
      ∈, ∉, set, dom, range, __⁺, __*, __⁻__, ×,
      ∪, ∩, ∘, -:R→R, ⁻¹, ⊆, ⊇, ⊂, ⊃,
      domRestrict, image, rangeRestrict

SetToRelation: trait
   % Map a (finitely generated) set
   % to the relation that represents it.
   assumes SetBasics, RelationBasics
   introduces
      relation: C → R
   asserts
      ∀ e: E, s: C
        relation({ }) == ⊥;
        relation(insert(e, s)) == [e, e] ∪ relation(s)
   implies
      ∀ e: E, s: C
        e ∈ s == e ⟨ relation(s) ⟩ e
      converts relation
```

The predicates in the next trait are closely related to the theories defined in Section A.11, but they define the properties of relations treated as values, whereas Section A.11 defines properties of relations treated as operators. This duplication is a price of not using a higher-order logic in LSL.

```
RelationPredicates: trait
  % Tests for useful properties
  % of individual relations.
  assumes
    RelationBasics,
    RelationOps
  introduces
    antisymmetric, asymmetric, equivalence,
        functional, irreflexive, oneToOne, reflexive,
        symmetric, total, transitive: R → Bool
    into, onto: R, R → Bool
  asserts
    ∀ r, r1, r2: R
      antisymmetric(r) == (r ∩ (r⁻¹)) ⊆ I;
      asymmetric(r) == r ∩ (r⁻¹) = ⊥;
      equivalence(r) ==
        reflexive(r) ∧ symmetric(r) ∧ transitive(r);
      functional(r) == ((r⁻¹) ∘ r) ⊆ I;
      irreflexive(r) == r ∩ I = ⊥;
      oneToOne(r) == r ∘ (r⁻¹) = I;
      reflexive(r) == I ⊆ r;
      symmetric(r) == r = r⁻¹;
      total(r) == dom(r) = I;
      transitive(r) == r = r⁺;
      into(r1, r2) == range(r1) ⊆ set(r2);
      onto(r1, r2) == set(r2) ⊆ range(r1);
  implies converts
    antisymmetric, asymmetric, equivalence,
    functional, irreflexive, oneToOne, reflexive,
    symmetric, total, transitive, into, onto
```

A.9   Graph theory

```
Graph (N, G): trait
   % n1 ( g ) n2 means that there is
   % an edge from n1 to n2 in g
   includes Relation (N for E, G for R)
   introduces
      nodes, undirected: G → G
      isPath: N, N, G → Bool
      stronglyConnected, weaklyConnected: G → Bool
   asserts ∀ n1, n2: N, g: G
      undirected(g) == g ∪ (g⁻¹);
      nodes(g) == dom(g) ∪ range(g);
      isPath(n1, n2, g) == n1 ( g* ) n2;
      stronglyConnected(g) == g* = nodes(g) × nodes(g);
      weaklyConnected(g) ==
         stronglyConnected(undirected(g))
   implies
      ∀ n1, n2: N, g: G
         (stronglyConnected(g) ∧ n1 ∈ nodes(g)
               ∧ n2 ∈ nodes(g))
            ⇒ isPath(n1, n2, g)
```

A.10   Properties of single operators

```
Associative (o, T): trait
  introduces __ o __ : T, T → T
  asserts ∀ x, y, z: T
    (x o y) o z == x o (y o z)

Commutative (o, T, Range): trait
  introduces __ o __: T, T → Range
  asserts ∀ x, y: T
    x o y == y o x

AC (o, T): trait
  introduces __ o __ : T, T → T
  asserts ∀ x, y, z: T
    (x o y) o z == x o (y o z);
    x o y == y o x
  implies
    Associative,
    Commutative (T for Range)

Idempotent (op, T): trait
  introduces op: T → T
  asserts ∀ x: T
    op(op(x)) == op(x)

Involutive (op, T): trait
  introduces op: T → T
  asserts ∀ x: T
    op(op(x)) == x
```

A.11   Properties of relational operators

Compare with `RelationPredicates`, **page 189**

```
Antisymmetric (◊): trait
   introduces __ ◊ __: T, T → Bool
   asserts ∀ x, y: T
     (x ◊ y ∧ y ◊ x) ⇒ x = y

Asymmetric (◊): trait
   introduces __ ◊ __: T, T → Bool
   asserts ∀ x, y: T
     x ◊ y ⇒ ¬ (y ◊ x)

Functional (◊): trait
   introduces __ ◊ __: T, T → Bool
   asserts ∀ x, y, z: T
     (x ◊ y ∧ x ◊ z) ⇒ y = z;

Irreflexive (◊): trait
   introduces __ ◊ __: T, T → Bool
   asserts ∀ x: T
     ¬ (x ◊ x)

OneToOne (◊): trait
   introduces __ ◊ __: T, T → Bool
   asserts ∀ x, y, z: T
     (x ◊ y ∧ x ◊ z) ⇒ y = z;
     (x ◊ z ∧ y ◊ z) ⇒ x = y;

Reflexive (◊): trait
   introduces __ ◊ __: T, T → Bool
   asserts ∀ x: T
     x ◊ x

Symmetric (◊): trait
   introduces __ ◊ __: T, T → Bool
   asserts ∀ x, y: T
     x ◊ y == y ◊ x
   implies Commutative (◊ for o, Bool for Range)
```

```
Transitive (◇): trait
  introduces __ ◇ __: T, T → Bool
  asserts ∀ x, y, z: T
    (x ◇ y ∧ y ◇ z) ⇒ x ◇ z

Equivalence: trait
  includes
    (Reflexive, Symmetric, Transitive)(≡ for ◇)

Equality (T): trait
  % This trait is given for documentation only.
  % It is implicit in LSL.
  introduces __ = __, __ ≠ __: T, T → Bool
  asserts
    T partitioned by =
   ∀ x, y, z: T
     x = x;
     x = y == y = x;
     (x = y ∧ y = z) ⇒ x = z;
     x ≠ y == ¬ (x = y)
  implies Equivalence (= for ≡)
```

A.12   Orderings

PARTIAL AND TOTAL ORDERS

```
IsPO (≤, T): trait
  % ≤ is a partial order on T
  introduces __≤__: T, T → Bool
  asserts ∀ x, y, z: T
    x ≤ x;
    (x ≤ y ∧ y ≤ z) ⇒ x ≤ z;
    x ≤ y ∧ y ≤ x == x = y
  implies
    Antisymmetric (≤),
    PreOrder,
    Reflexive (≤),
    Transitive (≤)
    T partitioned by ≤

PartialOrder (T): trait
  includes IsPO, DerivedOrders
  implies
    PartialOrder (> for <, < for >,
                  ≥ for ≤, ≤ for ≥),
    StrictPartialOrder (<, T)

IsTO (≤, T): trait
  % ≤ is a total order on T
  introduces __≤__: T, T → Bool
  asserts ∀ x, y, z: T
    x ≤ x;
    (x ≤ y ∧ y ≤ z) ⇒ x ≤ z;
    x ≤ y ∧ y ≤ x == x = y;
    x ≤ y ∨ y ≤ x
  implies IsPO, TotalPreOrder

TotalOrder (T): trait
  includes IsTO, DerivedOrders
  implies
    PartialOrder,
    StrictTotalOrder (<, T),
    TotalOrder (≥ for ≤, ≤ for ≥,
                > for <, < for >)
    T partitioned by <
```

ASSUMPTIONS AND IMPLICATIONS

```
PreOrder (≤, T): trait
  includes Reflexive (≤), Transitive (≤)
  implies ∀ x, y, z: T
    x ≤ x;
    (x ≤ y ∧ y ≤ z) ⇒ x ≤ z

TotalPreOrder (≤, T): trait
  includes PreOrder
  asserts ∀ x, y: T
    x ≤ y ∨ y ≤ x

StrictPartialOrder (<, T): trait
  includes Irreflexive (<), Transitive (<)
  implies
    Asymmetric (<)
    ∀ x, y, z: T
      ¬ (x < x);
      (x < y ∧ y < z) ⇒ x < z

StrictTotalOrder (<, T): trait
  includes StrictPartialOrder
  asserts ∀ x, y: T
    x < y ∨ y < x ∨ x = y
```

OPERATOR DEFINITIONS

```
DerivedOrders (T): trait
  % Define any three of the comparison operators,
  % given the fourth
  introduces
    __≤__, __≥__, __<__, __>__: T, T → Bool
  asserts ∀ x, y: T
    x ≤ y == x < y ∨ x = y;
    x < y == x ≤ y ∧ ¬ (x = y);
    x ≥ y == y ≤ x;
    x > y == y < x
  implies
    converts ≥, <, >
    converts ≤, <, >
    converts ≤, ≥, >
    converts ≤, ≥, <
```

```
MinMax (T): trait
  assumes TotalOrder
  introduces
    min, max: T, T → T
  asserts ∀ x, y: T
    min(x, y) == if x ≤ y then x else y;
    max(x, y) == if x ≥ y then x else y
  implies
    AC (min, T),
    AC (max, T)
    converts min, max

LexicographicOrder (E, C): trait
  % "Dictionary" order on C
  assumes
    Container,
    StrictTotalOrder (<, E)
  includes DerivedOrders (C)
  asserts ∀ c1, c2: C
    c1 < c2 ==
        c2 ≠ empty
          ∧ (c1 = empty
             ∨ (if head(c1) = head(c2)
                then tail(c1) < tail(c2)
                else head(c1) < head(c2)))
  implies
    TotalOrder (C)
    converts ≤:C,C→Bool, ≥:C,C→Bool,
      <:C,C→Bool, >:C,C→Bool
```

A.13   Lattice theory

```
GreatestLowerBound (T): trait
  introduces
    __ ≤ __: T, T → Bool
    __ ⊓ __: T, T → T
  asserts ∀ x, y, z: T
    (x ⊓ y) ≤ x;
    (x ⊓ y) ≤ y;
    (z ≤ x ∧ z ≤ y) ⇒ z ≤ (x ⊓ y)

Semilattice (T): trait
  assumes PartialOrder
  includes GreatestLowerBound
  introduces
    ⊥: → T
    __⊔ __: T, T → T
  asserts ∀ x, y, z: T
    ⊥ ≤ x;
    x ⊔ y == y ⊔ x;
    x ⊓ y == y ⊓ x;
    x ≤ (x ⊔ y);
    (x ≤ z ∧ y ≤ z) ⇒ (x ⊔ y) ≤ z
  implies
    AbelianMonoid (⊔ for o, ⊥ for unit),
    AbelianSemigroup (⊓ for o)

Lattice (T): trait
  assumes PartialOrder
  includes Semilattice
  introduces ⊤: → T
  asserts ∀ x: T
    x ≤ ⊤
  implies
    Lattice (⊔ for ⊓, ⊓ for ⊔, ⊤ for ⊥, ⊥ for ⊤,
             ≤ for ≥, ≥ for ≤, < for >, > for <)
```

A.14 Group theory

```
Semigroup: trait
  introduces __o __: T, T → T
  asserts ∀ x, y, z: T
    (x o y) o z == x o (y o z)
  implies Associative

LeftIdentity: trait
  introduces
    __ o __: T, T → T
    unit: → T
  asserts ∀ x: T
    unit o x == x

RightIdentity: trait
  introduces
    __ o __: T, T → T
    unit: → T
  asserts ∀ x: T
    x o unit == x

Identity: trait
  includes LeftIdentity, RightIdentity

Monoid: trait
  introduces
    __o __: T, T → T
    unit: → T
  asserts ∀ x, y, z: T
    (x o y) o z == x o (y o z);
    unit o x == x;
    x o unit == x
  implies Semigroup, Identity

LeftInverse: trait
  assumes LeftIdentity
  introduces __ ⁻¹: T → T
  asserts ∀ x: T
    (x⁻¹) o x == unit
```

```
RightInverse: trait
  assumes RightIdentity
  introduces __⁻¹: T → T
  asserts ∀ x: T
    x o (x⁻¹) == unit

Inverse: trait
  assumes Identity, Semigroup
  includes LeftInverse, RightInverse
  implies
    Involutive (__⁻¹ for op)
    ∀ x, y: T
      unit⁻¹ == unit;
      (x o y)⁻¹ == (y⁻¹) o (x⁻¹)

Group: trait
  introduces
    __o __: T, T → T
    unit: → T
    __⁻¹: T → T
  asserts ∀ x, y, z: T
    (x o y) o z == x o (y o z);
    unit o x == x;
    (x⁻¹) o x == unit;
  implies Monoid, Inverse

Abelian: trait
  introduces __ o __: T, T → T
  asserts ∀ x, y: T
    x o y == y o x
  implies Commutative (T for Range)

AbelianSemigroup: trait
  includes Abelian, Semigroup
  implies AC

AbelianMonoid: trait
  includes Abelian, Monoid

AbelianGroup: trait
  includes Abelian, Group
```

```
LeftDistributive (+, *, T): trait
   introduces
      __+__, __*__: T, T → T
   asserts ∀ x, y, z: T
      x * (y + z) == (x * y) + (x * z)

RightDistributive (+, *, T): trait
   introduces
      __+__, __*__: T, T → T
   asserts ∀ x, y, z: T
      (y + z) * x == (y * x) + (z * x)

Distributive (+, *, T): trait
   includes LeftDistributive, RightDistributive

Ring: trait
   includes
      AbelianGroup (+ for o, 0 for unit, -__ for ⁻¹),
      Semigroup (* for o),
      Distributive (+, *, T)

RingWithUnit: trait
   includes Ring, Monoid (* for o, 1 for unit)

Field: trait
   includes
      RingWithUnit,
      Abelian (* for o)
   introduces __⁻¹: T → T
   asserts ∀ x: T
      x ≠ 0 ⇒ x * (x⁻¹) = 1
```

A.15   Number theory

This section presents a series of traits dealing with operators on whole numbers. The following section deals with operators on rational and floating point numbers.

DATA TYPES

```
Natural (N): trait
  % The usual operators on the natural numbers,
  % starting from 0.
  includes
    ArithOps (N),
    DecimalLiterals,
    Exponentiation (N),
    MinMax (N),
    TotalOrder (N)
  introduces
    __ ⊖ __: N, N → N
  asserts
    N generated by 0, succ
    ∀ x, y: N
      succ(x) ≠ 0;
      succ(x) = succ(y) == x = y;
      x < succ(x);
      0 ⊖ x == 0;
      x ⊖ 0 == x;
      succ(x) ⊖ succ(y) == x ⊖ y
  implies
    NaturalOrder
    N generated by 0, 1, +
    ∀ x, y: N
      x ⊖ x == 0;
      x ≤ y == x ⊖ y = 0
    converts 1:→N, +, ⊖, *, div, mod,
        **, min, max, ≤, ≥, <, >
      exempting ∀ x: N
        div(x, 0), mod(x, 0)
```

```
Positive (P): trait
   % Basic operators on natural numbers,
   % starting from 1
   includes DecimalLiterals (P for N), TotalOrder (P)
   introduces
      1: → P
      succ: P → P
      __+__, __*__: P, P → P
   asserts
      P generated by 1, succ
      ∀ x, y: P
         x + 1 == succ(x);
         x + succ(y) == succ(x + y);
         x*1 == x;
         x*succ(y) == x + (x*y);
         x < succ(x)
   implies
      NaturalOrder (P for N, 1 for 0)
      P generated by 1, +
      converts +, *, ≤, ≥, <, >
```

```
IntCycle (first, last, N): trait
  % A finite subrange of the integers that includes 0,
  % and wraps at succ(last)
  includes
    ArithOps (N),
    DecimalLiterals,
    MinMax (N),
    TotalOrder (N)
  introduces
    first, last: → N
    pred, -__, abs: N → N
    __-__: N, N → N
  asserts
    N generated by 0, succ
    ∀ x, y: N
      succ(last) == first;
      pred(succ(x)) == x;
      succ(pred(x)) == x;
      -0 == 0;
      -succ(x) == pred(-x);
      abs(x) == if x < 0 then -x else x;
      x - y == x + (-y);
      x ≠ last ⇒ x < succ(x)
  implies
    Distributive (+, *, N),
    RingWithUnit (N for T)
    N generated by 0, pred
    ∀ x: N
      pred(first) == last;
      first ≤ x;
      x ≤ last;
      -(-x) == x
    converts
      pred, -__:N→N, abs, __-__:N,N→N,
      1:→N, +, *, max, min, ≤, ≥, <, >

SignedInt (maxSigned, N): trait
  % Typical machine arithmetic, signed complement.
  includes IntCycle (minSigned, maxSigned, N)
  asserts equations
    succ(minSigned) == -maxSigned
  implies equations
    minSigned + maxSigned == -1;
    abs(minSigned) == minSigned
```

```
UnsignedInt (maxUnsigned, N): trait
  % Typical machine arithmetic, unsigned.
  includes IntCycle (0, maxUnsigned, N)
```

ASSUMPTIONS AND IMPLICATIONS

Enumerable requires only that each value of sort N must be reachable by applying succ to 0 a finite number of times. Infinite requires that the values yielded by succ are all distinct. The inclusion of TotalOrder in NaturalOrder ensures that succ(x) is always greater than x, and hence that there are infinitely many distinct values of sort N.

```
Enumerable (N): trait
  introduces
    0: → N
    succ: N → N
  asserts
    N generated by 0, succ

Infinite (N): trait
  introduces
    0: → N
    succ: N → N
  asserts ∀ x, y: N
    succ(x) ≠ 0;
    succ(x) = succ(y) == x = y

NaturalOrder (N): trait
  % The natural numbers with an ordering
  includes
    Enumerable (N),
    TotalOrder (N)
  asserts ∀ x: N
    x < succ(x)
  implies
    Infinite (N)
    ∀ x, y: N
      0 ≤ x;
      x < succ(y) == x ≤ y;
      succ(x) < succ(y) == x < y
    converts ≤, ≥, <, >
```

OPERATOR DEFINITIONS

Addition (N): trait
```
  % Define the operator + in terms of 0 and succ
  includes AbelianMonoid(+ for o, 0 for unit, N for T)
  introduces
    0: → N
    succ: N → N
    __+__: N, N → N
  asserts ∀ x, y: N
    x + 0 == x;
    x + succ(y) == succ(x + y)
```

Multiplication (N): trait
```
  % Define the operator * in terms of 0, succ, and +
  includes
    AbelianMonoid (* for o, 1 for unit, N for T),
    Addition (N)
  introduces
    1: → N
    __*__: N, N → N
  asserts ∀ x, y: N
    1 == succ(0);
    x * 0 == 0;
    x * succ(y) == x + (x * y)
```

ArithOps (N): trait
```
  % Defines operators div and mod relative to + and *
  % for positive denominators
  assumes TotalOrder (N)
  includes Multiplication (N)
  introduces
    div, mod: N, N → N
  asserts ∀ x, y: N
    y > 0
      ⇒ (0 ≤ mod(x, y)
           ∧ mod(x, y) < y
           ∧ (mod(x, y) + (div(x, y) * y)) = x)
```

```
Exponentiation (T): trait
  % Repeatedly apply an infix * operator
  assumes
    Enumerable (N),
    Monoid (* for o, 1 for unit)
  introduces __**__: T, N → T
  asserts ∀ x: T, y: N
    x**0 == 1;
    x**succ(y) == x * (x**y)
  implies ∀ x: T
    x**succ(0) == x

IntegerAndNatural (Int, N): trait
  % Conversions between Int's and N's
  includes
    Integer (Int),
    Natural (N)
  introduces
    int: N → Int
    nat: Int → N
  asserts ∀ n: N
    int(0) == 0;
    int(succ(n)) == succ(int(n));
    nat(int(n)) == n

IntegerAndPositive (Int, P): trait
  % Conversions between Int's and P's
  includes
    Integer (Int),
    Positive (P)
  introduces
    int: P → Int
    pos: Int → P
  asserts ∀ p: P
    int(1) == 1;
    int(succ(p)) == succ(int(p));
    pos(int(p)) == p
```

A.16   Floating point arithmetic

The trait Rational provides enough of a theory of rational arithmetic to specify the properties of floating point arithmetic.

```
Rational: trait
  % For use in the trait FloatingPoint.
  includes
     Exponentiation (Q for T, P for N),
     IntegerAndPositive (Int, P),
     MinMax (Q),
     TotalOrder (Q)
  introduces
     __/__: Int, P → Q
     0, 1: → Q
     -__, __⁻¹, abs: Q → Q
     __+__, __*__, __-__, __/__: Q, Q → Q
  asserts
     Q generated by __/__: Int,P→Q
     ∀ i, i1, i2: Int, p, p1, p2, p3: P, q, q1, q2: Q
        0/p == 0;
        int(p)/p == 1;
        i1/p1 = i2/p2 == i1 * int(p2) = i2 * int(p1);
        -(i/p) == (-i)/p;
        (int(p1)/p2)⁻¹ == int(p2)/p1;
        (-q)⁻¹ == -(q⁻¹);
        abs(i/p) == abs(i)/p;
        (i1/p) + (i2/p) == (i1 + i2)/p;
        (i1/p1) * (i2/p2) == (i1 * i2)/(p1 * p2);
        q1 - q2 == q1 + (-q2);
        q1/q2 == q1 * (q2⁻¹);
        (i1/p) < (i2/p) == i1 < i2
  implies
     AC (+, Q),
     AC (*, Q),
     Field (Q for T)
     ∀ i, i1, i2: Int, p, p1, p2, p3: P, q: Q
        q + 0 == q;
        -q == 0 - q;
        (i1/p) - (i2/p) == (i1 - i2)/p;
        q * 0 == 0;
        q * 1 == q;
        q⁻¹ == 1/q;
        (i/p1)/(int(p2)/p3) == (i * int(p3))/(p1 * p2)
```

```
converts
     0:→Q, 1:→Q, -:Q →Q, ⁻¹, abs:Q →Q,
     +:Q,Q→Q, -:Q,Q→Q, *:Q,Q→Q, /:Q,Q→Q,
     **:Q,P→Q, min:Q,Q→Q, max:Q,Q→Q,
     <:Q,Q→Bool, >:Q,Q→Bool,
     ≤:Q,Q→Bool, ≥:Q,Q→Bool
exempting 0⁻¹
```

The following traits define a theory of floating point arithmetic that is weak enough to be satisfied by many floating point implementations, yet strong enough to allow reasoning about floating point arithmetic. Careful analysis of any particular floating point system should lead to tighter bounds on the errors due to inexact arithmetic, and might even lead to some useful identities, such as $(f_1 + f_2) + f_3 = f_1 + (f_2 + f_3)$.

The basic idea is this: Every floating point number exactly represents some rational number, returned by the operator `rational`. Each floating point operator approximates a corresponding rational operator, but cannot always be be exact. The exact answer may not even be representable. Furthermore, floating point arithmetic does not generally guarantee to produce even the closest representable value. So each floating point operator may introduce an error that depends on:

- the magnitude of the operand(s),

- the magnitude of the exact and approximate results,

- properties of the floating point representation used.

Three parameters characterize the representation: `smallest` and `largest` denote the least and the greatest representable positive values, respectively, and `gap`, the largest relative difference between any pair of consecutive representable positive values. `FPAssumptions` specifies relations that must hold among these parameters and the operator `rational` (which converts floating point numbers to their exact rational values) in order for `FloatingPoint` to characterize a valid floating point number system.

```
FPAssumptions (smallest, largest,
                gap, rational): trait
  includes Rational
  introduces
    smallest, largest, gap: → Q
    rational: F → Q
    float: Q → F
    0, 1: → F
  asserts ∀ f: F
    smallest > 0;
    largest > smallest;
    rational(0) == 0;
    rational(1) == 1;
    rational(f) ≠ 0 ⇒ abs(rational(f)) ≥ smallest;
    rational(f) ≤ largest;
    gap > 0;
    float(rational(f)) == f;
```

The predicate `approx(f, q, t)` compares the result f of a floating point operation to the exact rational value q of that operation; the predicate is true if the result is "close enough" to the exact value (i.e., within a tolerance t), or if the exact value is too big to be represented.

We have not axiomatized the properties of the IEEE standard's non-numeric floating point values (NaN's). We leave that as an exercise for numerical analysts, in the expectation that an accurate characterization is separable from the numerical properties. It might be more complex than anything we have specified in this handbook.

```
FloatingPoint (smallest, largest,
               gap, rational): trait
  assumes FPAssumptions
  includes
    Rational,
    TotalOrder (F)
  introduces
    mag: F → Q
    approx: F, Q, Q → Bool
    -__, abs, __⁻¹: F → F
    __+__, __*__, __-__, __/__: F, F → F
  asserts
    F generated by float
    ∀ f, f1, f2: F, q, t: Q
      f1 ≤ f2 == rational(f1) ≤ rational(f2);
      mag(f) == abs(rational(f));
      approx(f, q, t) ==
        abs(q) ≤ largest
          ⇒ abs(rational(f) - q)
            ≤ (smallest +
                (gap*(mag(f) + abs(q) + t)));
      approx(-f, -rational(f), 0);
      f ≠ 0 ⇒ approx(f⁻¹, rational(f)⁻¹, 0);
      approx(abs(f), mag(f), 0);
      approx(f1 + f2, rational(f1) + rational(f2),
              mag(f1) + mag(f2));
      approx(f1 * f2, rational(f1) * rational(f2), 0);
      approx(f1 - f2, rational(f1) - rational(f2),
              mag(f1) + mag(f2));
      f2 ≠ 0
        ⇒ approx(f1/f2, rational(f1)/rational(f2), 0)
```

Appendix B

Implementations of Example LCL Interfaces

This appendix contains the implementations of the interfaces erc, empset, and dbase specified in Chapter 5. We present them here not because they are intrinsically interesting, but for completeness.

ERC.H

```
#if !defined(ERC_H)
#define ERC_H

#include "eref.h"

typedef struct _elem{eref val; struct _elem *next;} ercElem;
typedef ercElem *ercList;
typedef struct {ercList vals; int size;} ercInfo;
typedef ercInfo *erc;
typedef ercList *ercIter;

#include "erc.lh"

#define erc_size(c) ((c)->size)
#define erc_choose(c) ((c->vals)->val)
#define erc_initMod( )\
        do {bool_initMod(); employee_initMod();\
        eref_initMod();} while (0)
#define erc_iterFinal(it) (free(it))
#define erc_iterReturn(it, result)\
    do {erc_iterFinal(it); return result;} while (0)
#define for_ercElems(er, it, c)\
   for(er = erc_yield(it = erc_iterStart(c));\
       !eref_equal(er, erefNIL);\
       er = erc_yield(it))
#endif
```

ERC.C

```c
#include "erc.h"

erc erc_create(void) {
  erc c;
  c = (erc) malloc(sizeof(ercInfo));
  if (c == 0) {
    printf("Malloc returned null in erc_create\n");
    exit(1);
  }
  c->vals = 0;
  c->size = 0;
  return c;
}

void erc_clear(erc c) {
  ercList elem;
  ercList next;
  for (elem = c->vals; elem != 0; elem = next) {
    next = elem->next;
    free(elem);
  }
  c->vals = 0;
  c->size = 0;
}

void erc_final(erc c) {
  erc_clear(c);
  free(c);
}

bool erc_member(eref er, erc c) {
  ercList tmpc;
    for (tmpc = c->vals; tmpc != 0; tmpc = tmpc->next)
    if (tmpc->val == er) return TRUE;
  return FALSE;
}

void erc_insert(erc c, eref er) {
  ercList newElem;
  newElem = (ercElem *) malloc(sizeof(ercElem));
  if (newElem == 0) {
    printf("Malloc returned null in erc_insert\n");
    exit(1);
  }
  newElem->val = er;
  newElem->next = c->vals;
  c->vals = newElem;
  c->size++;
}

bool erc_delete(erc c, eref er) {
  ercList elem;
```

```
   ercList prev;

   for (prev = 0, elem = c->vals;
        elem != 0;
        prev = elem, elem = elem->next) {
     if (elem->val == er) {
       if (prev == 0)
         c->vals = elem->next;
         else {prev->next = elem->next;}
       free(elem);
       c->size--;
       return TRUE;
     }
   }
   return FALSE;
}

ercIter erc_iterStart(erc c) {
   ercIter result;
   result = (ercIter) malloc(sizeof(ercList));
   if (result == 0) {
     printf("Malloc returned null in erc_iterStart\n");
     exit(1);
   }
   *result = c->vals;
   return result;
}

eref erc_yield(ercIter it) {
   eref result;
   if (*it == 0) {
     return erefNIL;
     free(it);
   }
   result = (*it)->val;
   *(it) = (*it)->next;
   return result;
}

void erc_join(erc c1, erc c2) {
   ercList tmpc;
   for(tmpc = c2->vals; tmpc != 0; tmpc = tmpc->next)
     erc_insert(c1, tmpc->val);
}

char * erc_sprint(erc c) {
   int len;
   eref er;
   ercIter it;
   char *result;
   result = (char*)malloc(erc_size(c)
                          * (employeePrintSize+1)+1);
   if (result == 0) {
```

```
      printf("Malloc returned null in erc_sprint\n");
      exit(1);
   }
   len = 0;
   for_ercElems (er, it, c) {
     employee_sprint(&(result[len]), eref_get(er));
     len += employeePrintSize;
     result[len++] = '\n';
   }
   result[len] = '\0';
   return result;
}
```

EMPSET.H

```
#if !defined(EMPSET_H)
#define EMPSET_H

#include "eref.h"
#include "erc.h"
#include "ereftab.h"

typedef erc empset;

ereftab known;

/*
  Abstraction function, toEmpSet:
    e \in toEmpSet(s) ==
      exists er (count(er, s.val) = 1
        /\ getERef(known, e) = er)

  Rep invariant:
    forall s: empset
      (forall er: eref (count(er, s.val) <= 1)
      /\ s.activeIters = 0
      /\ forall er: eref (count(er, s.val) = 1
        => in(known, er)))
*/

#include "empset.lh"

#define empset_create()    (erc_create())
#define empset_final(s)    (erc_final(s))
#define empset_member(e, s)\
        (!(eref_equal(_empset_get(e, s), erefNIL)))
#define empset_size(es)    (erc_size(es))
#define empset_choose(es)  (eref_get(erc_choose(es)))
#define empset_sprint(es)  (erc_sprint(es))
#endif
```

EMPSET.C

```
#include "empset.h"

static bool initDone = FALSE;

eref _empset_get(employee e, erc s) {
  eref er;
  ercIter it;
  employee e1;
  for_ercElems(er, it, s) {
    e1 = eref_get(er);
    if (employee_equal(&e1, &e))
      erc_iterReturn(it, er);
  }
  return erefNIL;
}

void empset_clear(empset s) {
  erc_clear(s);
}

bool empset_insert(empset s, employee e) {
  eref er;
  if (!eref_equal(_empset_get(e, s), erefNIL)) return FALSE;
  empset_insertUnique(s, e);
  return TRUE;
}

void empset_insertUnique(empset s, employee e) {
  eref er;
  er = ereftab_lookup(e, known);
  if (eref_equal(er, erefNIL)) {
    er = eref_alloc( );
    eref_assign(er,e);
    ereftab_insert(known, e, er);
  }
  erc_insert(s, er);
}

bool empset_delete(empset s, employee e) {
  eref er;
  er = _empset_get(e, s);
  if (eref_equal(er, erefNIL)) return FALSE;
  return erc_delete(s, er);
}

empset empset_disjointUnion(empset s1, empset s2) {
  erc result;
  ercIter it;
  eref er;
  empset tmp;
  result = erc_create( );
  if (erc_size(s1) > erc_size(s2)) {
```

```
    tmp = s1;
    s1 = s2;
    s2 = tmp;
  }
  erc_join(result, s1);
  for_ercElems(er, it, s2)
    empset_insertUnique(result, eref_get(er));
  return result;
}

empset empset_union(empset s1, empset s2) {
  eref er;
  ercIter it;
  erc result;
  empset tmp;
  result = erc_create( );
  if (erc_size(s1) > erc_size(s2)) {
    tmp = s1;
    s1 = s2;
    s2 = tmp;
  }
  erc_join(result, s2);
  for_ercElems(er, it, s1)
  if (!empset_member(eref_get(er), s2))
    erc_insert(result, er);
  return result;
}

void empset_intersect(empset s1, empset s2) {
  eref er;
  ercIter it;
  erc toDelete;
  toDelete = erc_create( );
  for_ercElems(er, it, s1)
    if (!empset_member(eref_get(er), s2))
      erc_insert(toDelete, er);
  for_ercElems(er, it, toDelete)
    erc_delete(s1, er);
  erc_final(toDelete);
}

bool empset_subset(empset s1, empset s2) {
  employee e;
  eref er;
  ercIter it;

  for_ercElems(er, it, s1)
    if (!empset_member(eref_get(er), s2))
      erc_iterReturn(it, FALSE);
  return TRUE;
}

void empset_initMod(void) {
```

```
    if (initDone) return;
    bool_initMod();
    employee_initMod();
    eref_initMod();
    erc_initMod();
    ereftab_initMod();
    known = ereftab_create( );
    initDone = TRUE;
}
```

DBASE.H

```
#if !defined(DBASE_H)
#define DBASE_H

#include "eref.h"
#include "erc.h"

#include "dbase.lh"

#endif
```

DBASE.C

```
#include <strings.h>
#include "dbase.h"

#define firstERC mMGRS
#define lastERC fNON
#define numERCS (lastERC - firstERC + 1)

typedef enum {mMGRS, fMGRS, mNON, fNON} employeeKinds;

erc db[numERCS];

bool initDone = FALSE;

void db_initMod(void) {
    int i;
    if (initDone) return;
    bool_initMod();
    employee_initMod();
    eref_initMod();
    erc_initMod();
    empset_initMod();
    for (i = firstERC; i <= lastERC; i++)
        db[i] = erc_create( );
    initDone = TRUE;
}

eref _db_ercKeyGet(erc c, int key) {
    eref er;
```

```
    ercIter it;
    for_ercElems(er, it, c)
        if (eref_get(er).ssNum == key) erc_iterReturn(it, er);
    return erefNIL;
}


eref _db_keyGet(int key) {
    int i;
    eref er;
    for (i = firstERC; i <= lastERC; i++) {
        er = _db_ercKeyGet(db[i], key);
        if (!eref_equal(er, erefNIL)) return er;
    }
    return erefNIL;
}

int _db_addEmpls(erc c, int l, int h, empset s) {
    eref er;
    ercIter it;
    employee e;
    int numAdded;
    numAdded = 0;
    for_ercElems (er, it, c) {
        e = eref_get(er);
        if ((e.salary >= l) && (e.salary <= h)) {
            empset_insert(s, e);
            numAdded++;
        }
    }
    return numAdded;
}

db_status hire(employee e) {
    if (e.gen == gender_ANY) return genderERR;
    if (e.j == job_ANY) return jobERR;
    if (e.salary < 0) return salERR;
    if (!eref_equal(_db_keyGet(e.ssNum), erefNIL))
        return duplERR;
    uncheckedHire(e);
    return db_OK;
}

void uncheckedHire(employee e) {
    eref er;
    er = eref_alloc();
    eref_assign(er, e);
    if (e.gen == MALE)
        if (e.j == MGR)
            erc_insert(db[mMGRS], er);
            else erc_insert(db[mNON], er);
        else if (e.j == MGR)
            erc_insert(db[fMGRS], er);
```

```
         else erc_insert(db[fNON], er);
}

bool fire(int ssNum) {
  int i;
  eref er;
  ercIter it;
  for (i = firstERC; i <= lastERC; i++)
    for_ercElems(er, it, db[i])
      if (eref_get(er).ssNum == ssNum) {
        erc_iterFinal(it);
        erc_delete(db[i], er);
        return TRUE;
      }
  return FALSE;
}

bool promote(int ssNum) {
  eref er;
  employee e;
  gender g;
  g = MALE;
  er = _db_ercKeyGet(db[mNON], ssNum);
  if (eref_equal(er, erefNIL)) {
    er = _db_ercKeyGet(db[fNON], ssNum);
    if (eref_equal(er, erefNIL)) return FALSE;
    g = FEMALE;
    }
  e = eref_get(er);
  e.j = MGR;
  eref_assign(er, e);
  if (g == MALE) {
     erc_delete(db[mNON], er);
     erc_insert(db[mMGRS], er);
     }
  else {
     erc_delete(db[fNON], er);
     erc_insert(db[fMGRS], er);
     }
  return TRUE;
}

db_status setSalary(int ssNum, int sal) {
  eref er;
  employee e;
  if (sal < 0) return salERR;
  er = _db_keyGet(ssNum);
  if (eref_equal(er, erefNIL)) return missERR;
  e = eref_get(er);
  e.salary = sal;
  eref_assign(er, e);
  return db_OK;
}
```

```
int query(db_q q, empset s) {
  eref er;
  employee e;
  int numAdded;
  int l, h;
  int i;
  l = q.l;
  h = q.h;
  switch(q.g) {
    case gender_ANY:
      switch(q.j) {
        case job_ANY:
          numAdded = 0;
          for (i = firstERC; i <= lastERC; i++)
            numAdded += _db_addEmpls(db[i], l, h, s);
          return numAdded;
        case MGR:
          numAdded = _db_addEmpls(db[mMGRS], l, h, s);
          numAdded += _db_addEmpls(db[fMGRS], l, h, s);
          return numAdded;
        case NONMGR:
          numAdded = _db_addEmpls(db[mNON], l, h, s);
          numAdded += _db_addEmpls(db[fNON], l, h, s);
          return numAdded;
        }
    case MALE:
      switch(q.j) {
        case job_ANY:
          numAdded = _db_addEmpls(db[mMGRS], l, h, s);
          numAdded += _db_addEmpls(db[mNON], l, h, s);
          return numAdded;
        case MGR:
          return _db_addEmpls(db[mMGRS], l, h, s);
        case NONMGR:
          return _db_addEmpls(db[mNON], l, h, s);
}
    case FEMALE:
      switch(q.j) {
        case job_ANY:
          numAdded = _db_addEmpls(db[fMGRS], l, h, s);
          numAdded += _db_addEmpls(db[fNON], l, h, s);
          return numAdded;
        case MGR:
          return _db_addEmpls(db[fMGRS], l, h, s);
        case NONMGR:
          return _db_addEmpls(db[fNON], l, h, s);
}
    }
}

void db_print(void) {
  int i;
```

```
  char * printVal;
  printf("Employees:\n");
  for (i = firstERC; i <= lastERC; i++) {
     printVal = erc_sprint(db[i]);
     printf("%s", printVal);
     free(printVal);
   }
}
```

Appendix C

Lexical Forms and Initialization Files

The Larch languages were designed for use with an open-ended collection of programming languages, support tools, and input/output facilities, each of which may have its own lexical conventions and capabilities. To conform to local conventions and to exploit locally available capabilities, character and token classes are extensible and can be tailored for particular purposes by *initialization files*.

In this appendix we give the LSL and LCL initialization files used for the examples in this book. We also give the ISO Latin codes used for typing the special symbols appearing in specifications in this book.

The book was produced using LaTeX with a Larch style file. That allowed us to type specifications using the ISO Latin codes given here, and have them appear in the text as special symbols.

LCL init file

```
commentSym  //

opChar      ~!#$&?@|

selectSym

synonym     \and        /\
synonym     \or         \/
synonym     \implies    =>
synonym     \marker     __
synonym     \eq         ==
synonym     \neq        !=
synonym     \not        !
synonym     \not        not
synonym     \not        ~
synonym     \pre        ^
synonym     \post       '
synonym     \arrow      ->
synonym     \arrow      \ra
```

LSL init file

```
commentSym    %

idChar        '
opChar        ~!#$&?@|
singleChar    ;

openSym       [ { \< \langle
closeSym      ] } \> \rangle
selectSym     .

simpleId      \bot \top

synonym       \and          /\
synonym       \and          &
synonym       \or           \/
synonym       \or           |
synonym       \implies      =>
synonym       \not          !
synonym       \not          not
synonym       \not          ~
synonym       \eq           =
synonym       \neq          !=
synonym       \neq          ~=
synonym       \arrow        ->
synonym       \marker       ___
synonym       \equals       ==
synonym       \forall       forall
synonym       \eqsep        ;

% Following used for checking LCL

synonym       Bool          bool
synonym       Int           int
synonym       Int           signed_char
synonym       Int           unsigned_char
synonym       Int           short_int
synonym       Int           long_int
synonym       Int           unsigned_short_int
synonym       Int           unsigned_int
synonym       Int           unsigned_long_int
synonym       double        float
synonym       double        long_double
```

ISO Latin codes for special characters

→ is written as ->
≤ is written as <=
≥ is written as >=
≠ is written as ~=
¬ is written as ~
∨ is written as \/
∧ is written as /\
⇒ is written as =>
∀ is written as \forall
∃ is written as \exists
• is written as \any
* is written as *
⁺ is written as \+
⁻¹ is written as \inv
⟨ is written as \<
⟩ is written as \>
∈ is written as \in
∉ is written as \notin
∩ is written as \I
∪ is written as \U
⊂ is written as \subset
⊆ is written as \subseteq
⊃ is written as \supset
⊇ is written as \supseteq
⊣ is written as -|
⊢ is written as |-
‖ is written as ||
· is written as \cdot
∘ is written as \circ
⊣ is written as \precat
⊢ is written as \postcat
⊥ is written as \bot
⊤ is written as \top
⊓ is written as \glb
⊔ is written as \lub
⊖ is written as \ominus
◇ is written as \rel
× is written as \times

Appendix D

Further Information and Tools

This appendix contains a list of currently available Larch tools.

Readers interested in keeping up with new developments should subscribe to the electronic mailing list larch-interest@src.dec.com. This list is used for announcements and queries of general interest. Requests to be added to (or deleted from) this list, as well as more specialized queries, should be sent to larch-interest-request@src.dec.com.

All information in this section is current as of October 1992. An updated version will be kept online on the internet host gatekeeper.dec.com. It will be available for anonymous ftp as

```
/pub/DEC/Larch/Information.tex
```

1. **lsl**. Larch Shared Language Checker. Syntax and sort checks LSL specifications. Translates LSL into **lp** input. Contact: Stephen Garland, MIT.

2. **lcl**. Syntax and type checker for LCL. Interfaces with **lsl**. Contact: Stephen Garland, MIT.

3. **lm3**. Syntax and type checker for Modula-3 interface specifications written in LM3. Interfaces with **lsl**. Contact: Kevin Jones, DEC.

4. **lp**. Larch Prover. Proof checker for fragment of first-order logic with equality. Contact: Stephen Garland, MIT.

5. **gcil**. Generic Concurrent Interface Language (GCIL) Checker. Syntax and type checks GCIL specifications. Interfaces with **lsl**. Contact: Jeannette Wing, CMU.

6. **Penelope**. Verification tool for Larch/Ada specifications and Ada programs. Contact: M. Stillman, ORA.

7. **Larch/Smalltalk Browser**. Syntax and sort/type checker and browser for Larch/Smalltalk and LSL specifications. Contact: Gary Leavens, ISU.

CONTACT ADDRESSES

MIT/LCS

Dr. Stephen J. Garland
Massachusetts Institute of Technology
Laboratory for Computer Science
545 Technology Square
Cambridge, MA 02139, USA
Internet:`garland@lcs.mit.edu`

DEC/SRC

Dr. James J. Horning
Dr. Kevin D. Jones
Digital Equipment Corporation
Systems Research Center
130 Lytton Avenue
Palo Alto, CA 94301-1044, USA
Internet: `horning@src.dec.com`, `kjones@src.dec.com`

ISU/DCS

Professor Gary Leavens
229 Atanasoff Hall
Department of ComputerScience
Iowa State University
Ames, Iowa 50011-1040, USA
Internet: `leavens@cs.iastate.edu.`

ORA

M. Stillman
Odyssey Research Associates
301A Harris B. Dates Drive
Ithaca, NY 14850-1313, USA.

CMU/SCS

Professor Jeannette M. Wing
Carnegie Mellon University
School of Computer Science
Pittsburgh, PA 15213-3890, USA
Internet: `Jeannette.Wing@cs.cmu.edu`

Appendix E

Classified Bibliography

This bibliography was started by Jeannette Wing and augmented by Yang Meng Tan. It is available by anonymous `ftp` from Internet node `larch.lcs.mit.edu` as `/pub/larch-bib/larch-bib.tex`. Suggested additions for the online version should be sent to `ymtan@lcs.mit.edu`. Full citations for all references are given in the next section.

Papers about Larch

CURRENT WORK

Reports about the current status of several Larch-related projects are contained in [66].

LARCH LANGUAGES

Larch Interface Languages: generic [16, 53, 61, 88]; Ada [37]; C [26, 80]; C++ [60]; CLU [86]; ML [93]; Modula-3 [55, 56, 57]; Smalltalk [17]. Larch and other methods: [95].

LARCH TOOLS

LP, the Larch proof assistant: [30]; a beginner's strategy guide [81]; an extension [83]; [5, 11, 18, 19, 76, 84].
For LSL [7, 59]; for LCL [26]; for LM3 [57].

Example specifications

Apple MAC Toolbox: [13].
Avalon built-in classes, examples (queue, directory, counter): [92], [89], and [61].
Display: [43].
Finite element analysis library: [3, 1].
Garbage collection: [22].

IOStreams: [55].
Larch/Ada: [15, 37].
Library: [87].
Miro languages and editor: [94, 99].
Thread synchronization primitives: [6, 69].
Using specifications to search software libraries: [73].

Proofs using LP

Ada programs: [38]
Avalon queue example: [92, 35, 91].
Circuit examples: [18, 32, 78, 75, 79].
Mathematical Theorems: [65].
Temporal Logic of Actions: [25].

References

* Entries marked with an asterisk have been superseded by material in this book; they are included for historical reference only.

[1] J.W. Baugh, Jr. "Formal specification of engineering analysis programs," *Expert Systems for Scientific Computing*, E.N. Houstis, J.R. Rice, and R. Vichnevetsky (eds.), North-Holland, 1992.

[2] John W. Baugh, Jr. "Is engineering software amenable to formal specification?," in [66].

[3] J.W. Baugh, Jr., and D.R. Rehak. *Computational Abstractions for Finite Element Programming*, TR 89-182, Dept. of Civil Engineering, Carnegie Mellon University, 1989.

[4] Michel Bidoit. *Pluss, un langage pour le développement de spécifications algébriques modulaires*. Thèse d'Etat, Université de Paris-Sud, Orsay, May 1989.

[5] Michel Bidoit and Rolf Hennicker, "How to prove observational theorems with LP," in [66].

[6] A.D. Birrell, J.V. Guttag, J.J. Horning, and R. Levin. "Synchronization primitives for a multiprocessor: a formal specification." *Operating Systems Review* 21(5), Nov. 1987. Revised version in [69].

[7] Robert H. Bourdeau and Betty H.C. Cheng. "An Object-oriented Toolkit for Constructing Specification Editors," *Proc. COMPSAC'92: Computer Software and Applications Conf.*, Sept. 1992.

[8] Robert S. Boyer and J S. Moore. *A Computational Logic*, Academic Press, 1979.

[9] Robert S. Boyer and J S. Moore. *A Computational Logic Handbook*, Academic Press, 1988.

[10] Frederick P. Brooks, Jr. *The Mythical Man-Month: Essays on Software Engineering*, Addison-Wesley, 1975.

[11] Manfred Broy. *Experiences with Software Specification and Verification Using LP, the Larch Proof Assistant*, TR 93, DEC/SRC, Oct. 1992.

[12] R.M. Burstall and J.A. Goguen. "Semantics of CLEAR, a specification language," *Proc. Advanced Course on Abstract Software Specifications*, D. Bjorner (ed.), Springer-Verlag, LNCS 86, 1980.

[13] C.T. Burton, S.J. Cook, S. Gikas, J.R. Rowson, and S.T. Sommerville. "Specifying the Apple Macintosh toolbox event manager," *Formal Aspects of Computing* 1(2), 1989.

[14] Karl-Heinz Buth. "Using SOS definitions in term rewriting proofs," in [66].

[15] S.R. Cardenas and H. Oktaba. *Formal Specification in Larch Case Study: Text Manager. Interface Specification, Implementation, in Ada and Validation of Implementation*, TR 511, Instituto de Investigaciones en Matematicas Aplicadas y en Sistemas, Universidad Nacional Autonoma de Mexico, 1988.

[16] Jolly Chen. *The Larch/Generic Interface Language*, S.B. Thesis, Dept. of Electrical Engineering and Computer Science, MIT, 1989.

[17] Yoonsik Cheon. *Larch/Smalltalk: A Specification Language for Smalltalk*, M.Sc. Thesis, Iowa State University, 1991.

[18] Boutheina Chetali and Pierre Lescanne. "An exercise in LP: the proof of a nonrestoring division circuit," in [66].

[19] Christine Choppy and Michel Bidoit. "Integrating ASSPEGIQUE and LP," in [66].

[20] O.-J. Dahl, D.F. Langmyhr, and O. Owe. *Preliminary Report on the Specification and Programming Language ABEL*, Research Report 106, Institute of Informatics, University of Oslo, Norway, 1986.

[21] Ole-Johan Dahl. *Verifiable Programming*, Prentice Hall International Series in Computer Science, 1992.

[22] David L. Detlefs. *Concurrent, Atomic Garbage Collection*, Ph.D. Thesis, Dept. of Computer Science, Carnegie Mellon University, TR CS-90-177, Oct. 1990.

[23] H.-D. Ehrich. "Extensions and implementations of abstract data type specifications," *Proc. Mathematical Foundations of Computer Science*, Zakopane, Sept. 1978. Springer-Verlag, LNCS 64.

[24] H. Ehrig and B. Mahr. *Fundamentals of Algebraic Specification 1: Equations and Initial Semantics*, EATCS Monographs on Theoretical Computer Science, vol. 6, Springer-Verlag, 1985.

[25] Urban Engberg, Peter Grønning, and Leslie Lamport. "Mechanical verification of concurrent systems with TLA," *Proc. Workshop on Computer Aided Verification*, 1992. Revised version in [66].

[26] G. Feldman and J. Wild. "The DECspec project: tools for Larch/C," *Proc. Fifth Int. Workshop on Computer-Aided Software Engineering*, Montreal, Jul. 1992. Revised version in [66].

[27] Stephen J. Garland and John V. Guttag. "Inductive methods for reasoning about abstract data types," *Proc. 15th ACM Symp. Principles of Programming Languages*, Jan. 1988.

[28] Stephen J. Garland and John V. Guttag. "An overview of LP, the Larch Prover," *Proc. Third Intl. Conf. Rewriting Techniques and Applications*, Chapel Hill, 1989. Springer-Verlag, LNCS 355.

[29] * Stephen J. Garland and John V. Guttag. "Using LP to debug specifications," *Proc. IFIP Work. Conf. Programming Concepts and Methods*, Tiberias, Apr. 1990. North-Holland.

[30] Stephen J. Garland and John V. Guttag. *A Guide to LP, The Larch Prover*, TR 82, DEC/SRC, Dec. 1991.

[31] * Stephen J. Garland, John V. Guttag, and James J. Horning. "Debugging Larch Shared Language specifications," *IEEE Trans. Software Engineering* 16(9), Sept. 1990.

[32] S.J. Garland, J.V. Guttag, and J. Staunstrup. "Verification of VLSI circuits using LP," *Proc. IFIP Work. Conf. Fusion of Hardware Design and Verification*, North Holland, 1988.

[33] Narain Gehani and Andrew McGettrick (eds.). *Software Specification Techniques*, Addison-Wesley, 1986.

[34] J.A. Goguen, J.W. Thatcher, and E.G. Wagner. "An initial algebra approach to the specification, correctness and implementation of abstract data types," *Current Trends in Programming Methodology IV: Data Structuring*, R. Yeh (ed.), Prentice-Hall, 1978.

[35] C. Gong and J.M. Wing. *Raw Code, Specification, and Proofs of the Avalon Queue Example*, Carnegie Mellon University, TR CMU-CS-89-172, Aug. 1989.

[36] David Gries. *The Science of Programming*, Springer-Verlag, 1981.

[37] David Guaspari, Carla Marceau, and Wolfgang Polak. "Formal verification of Ada," *IEEE Trans. Software Engineering* 16(9), Sept. 1990.

[38] David Guaspari, Carla Marceau, and Wolfgang Polak. "Formal verification of Ada programs," in [66].

[39] John V. Guttag. "Dyadic specification and its Impact on reliability," in *Three Approaches to Reliable Software: Language Design Dyadic Specification, Complementary Semantics*, J.E. Donahue, J.D. Gannon, J.V. Guttag, and J.J. Horning, University of Toronto, TR CSRG-45, Dec. 1974.

[40] John V. Guttag. *The Specification and Application to Programming of Abstract Data Types*, Ph.D. Thesis, Dept. of Computer Science, University of Toronto, 1975.

[41] John Guttag. "Notes on type abstraction," *Proc. Conf. Specifications of Reliable Software*, 1979. Reprinted in [33].

[42] J.V. Guttag and J.J. Horning. "The Algebraic Specification of Abstract Data Types," *Acta Informatica* 10(1), 1978.

[43] John Guttag and J.J. Horning. "Formal Specification as a Design Tool," *Seventh ACM Symp. Principles of Programming Languages*, Las Vegas, Jan. 1980. Reprinted in [33].

[44] * J.V. Guttag and J.J. Horning. "An Introduction to the Larch Shared Language," *Proc. IFIP Ninth World Computer Congress*, Paris, Sept. 1983.

[45] * J.V. Guttag and J.J. Horning. "Report on the Larch Shared Language," *Science of Computer Programming* 6(2), Mar. 1986.

[46] * J.V. Guttag and J.J. Horning. "A Larch Shared Language Handbook," *Science of Computer Programming* 6(2) Mar. 1986.

[47] * J.V. Guttag and J.J. Horning. *LCL: A Larch Interface Language for C*, TR 74, DEC/SRC, Jul. 1991.

[48] * John V. Guttag and James J. Horning. "A Tutorial on Larch and LCL, a Larch/C Interface Language," *Proc. VDM91: Formal Software Development Methods*, S. Prehn and W.J. Toetenel (eds.), Delft, Oct. 1991. Springer-Verlag, LNCS 551.

[49] * John V. Guttag, James J. Horning, and Andrés Modet. *Report on the Larch Shared Language: Version 2.3*, TR 58, DEC/SRC, Apr. 1990.

[50] * John V. Guttag, James J. Horning, and Jeannette M. Wing. "The Larch Family of Specification Languages," *IEEE Software* 2(5), 1985.

[51] * J.V. Guttag, J.J. Horning, and J.M. Wing. *Larch in Five Easy Pieces*, TR 5, DEC/SRC, Jul. 1985.

[52] Samuel P. Harbison. *Modula-3*, Prentice Hall, 1992.

[53] David Hinman. *On the Design of Larch Interface Languages*, S.M. Thesis, Dept. of Electrical Engineering and Computer Science, MIT, Jan. 1987.

[54] * J.J. Horning. "Combining Algebraic and Predicative Specifications in Larch," *Proc. Intl. Joint Conf. on Theory and Practice of Software Development, TAPSOFT*, Berlin, Mar. 1985. Springer-Verlag, LNCS 186.

[55] Kevin D. Jones. *LM3: A Larch Interface Language for Modula-3: A Definition and Introduction: Version 1.0*, TR 72, DEC/SRC, Jun. 1991.

[56] Kevin D. Jones. "A semantics for a Larch/Modula-3 interface language," in [66].

[57] Kevin D. Jones. *LM3 Reference Manual*, (in preparation).

[58] Donald E. Knuth and Peter B. Bendix. "Simple word problems in universal algebras," *Computational Problems in Abstract Algebra*, John Leech (ed.), Pergamon Press, Oxford, 1970.

[59] Michael R. Laux, Robert H. Bourdeau, and Betty H.C. Cheng. *An Integrated Environment Supporting the Reuse of Formal Specifications*, Michigan State University, Dept. of Computer Science, TR MSU-CPS-ACS-70, Sept. 1992.

[60] Gary T. Leavens and Yoonsik Cheon. "Preliminary design of Larch/C++," in [66].

[61] Richard Allen Lerner. *Specifying Objects of Concurrent Systems*, Ph.D. Thesis, Dept. of Computer Science, Carnegie Mellon University, TR CS-91-131, May 1991.

[62] Pierre Lescanne. "Computer experiments with the REVE term rewriting system generator," *Proc. Tenth ACM Symp. Principles of Programming Languages*, 1983.

[63] Barbara Liskov and John Guttag. *Abstraction and Specification in Program Development*, MIT EECS Series, MIT Press, 1986.

[64] D.A. McAllester. *Ontic: A Knowledge Representation System for Mathematics*, MIT Press.

[65] U. Martin and T. Nipkow. "Automating Squiggol," *Proc. IFIP Work. Conf. Programming Concepts and Methods*, Tiberias, Apr. 1990. North-Holland.

[66] U. Martin and J.M. Wing. *Proc. First Intl. Workshop on Larch*, Dedham, Jul. 1992, Springer-Verlag.

[67] Niels Mellergaard and Jørgen A. Staunstrup. "Generating proof obligations for circuits," in [66].

[68] James H. Morris, Jr. "Types are Not Sets," *First ACM Symp. Principles of Programming Languages*, Boston, Oct. 1973.

[69] Greg Nelson (ed.). *Systems Programming with Modula-3*, Prentice Hall, 1991.

[70] D.L. Parnas. "Information distribution aspects of design methodology," *Proc. IFIP Congress 71*, Ljubljana, Aug. 1971.

[71] Laurence C. Paulson. *Logic and Computation: Interactive Proof with Cambridge LCF*, Cambridge University Press, 1987.

[72] Gerald E. Peterson and Mark E. Stickel. "Complete sets of reductions for some equational theories," *J. ACM* 28:2, Apr. 1981.

[73] Eugene J. Rollins and Jeannette M. Wing. "Specifications as search keys for software libraries," *Proc. Intl. Conf. Logic Programming*, Paris, Jun. 1991.

[74] Donald Sannella and Andrzej Tarlecki. "On observational equivalence and algebraic specification," *Proc. Intl. Joint Conf. Theory and Practice of Software Development, TAPSOFT*, Berlin, Mar. 1985. Springer-Verlag, LNCS 185.

[75] James B. Saxe, Stephen J. Garland, John V. Guttag, and James J. Horning. "Using Transformations and Verification in Circuit Design," in [66].

[76] E. A. Scott and K. J. Norrie. "Using LP to study the language PL," in [66].

[77] Joseph R. Shoenfield. *Mathematical Logic*, Addison-Wesley, 1967.

[78] J. Staunstrup, S.J. Garland, and J.V. Guttag. "Compositional verification of VLSI circuits," *Proc. Intl. Workshop on Automatic Verification Methods for Finite State Systems*, Grenoble, Jun. 1989, Springer-Verlag, LNCS 407.

[79] Jørgen Staunstrup, Stephen J. Garland, and John V. Guttag. "Mechanized verification of circuit descriptions using the Larch Prover," *Proc. IFIP Work. Conf. Theorem Provers in Circuit Design: Theory, Practice, and Experience*, Nijmegen, Jun. 1992. North-Holland.

[80] Yang Meng Tan. "Semantic analysis of Larch interface specifications," in [66].

[81] David S. Taylor. *A Beginner's Strategy Guide to the Larch Prover*, S.B. Thesis, Dept. of Electrical Engineering and Computer Science, MIT, May 1990.

[82] Mark T. Vandevoorde. "Optimizing programs with partial specifications," in [66].

[83] Mary A. Vogt. *Extension of the Larch Prover by a Method of Inference Using Linear Arithmetic*, S.B. Thesis, Dept. of Electrical Engineering and Computer Science, MIT, Sept. 1990.

[84] Frederic Voisin. "A new front-end for the Larch Prover," in [66].

[85] M. Wand. "Final algebra semantics and data type extensions," *Journal of Computer and System Sciences*, Aug. 1979.

[86] Jeannette Marie Wing. *A Two-Tiered Approach to Specifying Programs*, Ph.D. Thesis, Dept. of Electrical Engineering and Computer Science, MIT, TR MIT/LCS/TR-299, May 1983.

[87] Jeannette M. Wing. "A Larch specification of the library problem," *Proc. Fourth Int. Workshop on Software Specification and Design*, Monterey, Apr. 1987.

[88] Jeannette M. Wing. "Writing Larch Interface Language Specifications," *ACM Trans. Programming Languages and Systems* 9(1), Jan. 1987.

[89] J.M. Wing. "Specifying recoverable objects," *Proc. Sixth Annual Northwest Software Quality Conf.*, Portland, Sept. 1990.

[90] J.M. Wing. "Using Larch to Specify Avalon/C++ Objects," *Proc. Intl. Joint Conf. Theory and Practice of Software Development, TAPSOFT*, Barcelona, Mar. 1989. Springer-Verlag, LNCS 352. Revised version in [90]

[91] Jeannette M. Wing and Chun Gong. *Machine-Assisted Proofs of Properties of Avalon Programs*, Carnegie Mellon University, TR CMU-CS-89-171, Aug. 1989.

[92] Jeannette M. Wing and Chun Gong. "Experience with the Larch Prover," *Proc. ACM Intl. Workshop on Formal Methods in Software Development*, May 1990.

[93] J.M. Wing, Eugene Rollins, and Amy Moormann Zaremski. "Thoughts on a Larch/ML and a new application for LP," in [66].

[94] Jeannette M. Wing and Amy Moormann Zaremski. "A formal specification of a visual language editor," *Proc. Sixth Intl. Workshop on Software Specification and Design*, Como, Oct. 1991.

[95] Jeannette M. Wing and A. Moormann Zaremski. "Unintrusive ways to integrate formal specifications in practice," *Proc. VDM91: Formal Software Development Methods*, S. Prehn and W.J. Toetenel (eds.), Delft, Oct. 1991. Springer-Verlag, LNCS 551.

[96] M. Wirsing. *Algebraic Specification*, Technical Report MIP-8914, University of Passau, Germany, 1989.

[97] Katherine Anne Yelick. *Using Abstractions in Explicitly Parallel Programs*, Ph.D. Thesis, Dept. of Electrical Engineering and Computer Science, MIT, TR MIT/LCS/TR-507, Jul. 1990.

[98] Katherine A. Yelick and Stephen J. Garland. "A parallel completion procedure for term rewriting systems," *Proc. 11th Intl. Conf. Automated Deduction*, Saratoga Springs, Jun. 1992. Springer-Verlag, LNCS 607.

[99] Amy Moormann Zaremski. *A Larch Specification of the Miro Editor*, Carnegie Mellon University, TR CMU-CS-91-111, Feb. 1991.

Index

Texts and Monographs in Computer Science

(continued from page ii)

Edsger W. Dijkstra and Carel S. Scholten
Predicate Calculus and Program Semantics
1990. XII, 220 pages

W.H.J. Feijen, A.J.M. van Gasteren, D. Gries, and J. Misra, Eds.
Beauty Is Our Business: A Birthday Salute to Edsger W. Dijkstra
1990. XX, 453 pages, 21 illus.

P.A. Fejer and D.A. Simovici
**Mathematical Foundations of Computer Science, Volume I:
Sets, Relations, and Induction**
1990. X, 425 pages, 36 illus.

Melvin Fitting
First-Order Logic and Automated Theorem Proving
1990. XIV, 242 pages, 26 illus.

Nissim Francez
Fairness
1986. XIII, 295 pages, 147 illus.

R.T. Gregory and E.V. Krishnamurthy
Methods and Applications of Error-Free Computation
1984. XII, 194 pages, 1 illus.

David Gries, Ed.
Programming Methodology: A Collection of Articles by Members of IFIP WG2.3
1978. XIV, 437 pages, 68 illus.

David Gries
The Science of Programming
1981. XV, 366 pages

John V. Guttag and James J. Horning
Larch: Languages and Tools for Formal Specification
1993. XIII, 250 pages, 76 illus.

Micha Hofri
Probabilistic Analysis of Algorithms
1987. XV, 240 pages, 14 illus.

A.J. Kfoury, Robert N. Moll, and Michael A. Arbib
A Programming Approach to Computability
1982. VIII, 251 pages, 36 illus.

Texts and Monographs in Computer Science

(continued)

Dexter C. Kozen
The Design and Analysis of Algorithms
1992. X, 320 pages, 72 illus.

E.V. Krishnamurthy
Error-Free Polynomial Matrix Computations
1985. XV, 154 pages

David Luckham
Programming with Specifications: An Introduction to ANNA, A Language for Specifying Ada Programs
1990. XVI, 418 pages, 20 illus

Ernest G. Manes and Michael A. Arbib
Algebraic Approaches to Program Semantics
1986. XIII, 351 pages

Robert N. Moll, Michael A. Arbib, and A.J. Kfoury
An Introduction to Formal Language Theory
1988. X, 203 pages, 61 illus.

Helmut A. Partsch
Specification and Transformation of Programs
1990. XIII, 493 pages, 44 illus.

Franco P. Preparata and Michael Ian Shamos
Computational Geometry: An Introduction
1988. XII, 390 pages, 231 illus.

Brian Randell, Ed.
The Origins of Digital Computers: Selected Papers, 3rd Edition
1982. XVI, 580 pages, 126 illus.

Thomas W. Reps and Tim Teitelbaum
The Synthesizer Generator: A System for Constructing Language-Based Editors
1989. XIII, 317 pages, 75 illus.

Thomas W. Reps and Tim Teitelbaum
The Synthesizer Generator Reference Manual, 3rd Edition
1989. XI, 171 pages, 79 illus.

Arto Salomaa and Matti Soittola
Automata-Theoretic Aspects of Formal Power Series
1978. X, 171 pages

Texts and Monographs in Computer Science

DATE DUE		L.-Brault
0 6 MARS 1995		
2 1 NOV. 1995		
2 1 OCT. 1995		
0 3 NOV. 1995		
2 1 NOV. 1995		
1 5 DEC. 1995		
1 9 DEC. 1995		
1 0 JUIN 1996		
2 3 AOUT 1996		
2 1 OCT. 1996		
RESERVE 2 5 SEP. 1997		

Bibliofiche 297B